U.S. DEPARTMENT OF HOUSING AND

Residential
Rehabilitation Inspection Guide

Residential Inspection

Residential Rehabilitation Inspection Guideline

**Prepared for the
U.S. Department of Housing and Urban Development
Office of Policy Development and Research**

by the National Institute of Building Sciences, Washington, D.C.
under Contract C-OPC-21204

February 2000

Residential Inspection

PATH (Partnership for Advancing Technology in Housing) is a new private/public effort to develop, demonstrate, and gain widespread acceptance for the "Next Generation" of American housing. Through the use of new and innovative technologies the goal of PATH is to improve the quality, durability, environmental efficiency, and affordability of tomorrow's houses.

Initiated at the request of the White House, PATH is managed and supported by the U.S. Department of Housing and Urban Development. In addition, all Federal Agencies that engage in housing research and technology development are PATH partners, including the Department of Energy, the Department of Commerce, the Environmental Protection Agency, and the Federal Emergency Management Agency. State and local governments and other participants in the public sector also are partners in PATH. Product manufacturers, home builders, insurance companies, and lenders represent private industry in the PATH partnership.

To learn more, please contact: PATH, Suite B133, 451 Seventh Street, S.W., Washington, D.C. 20410; fax 202-708-4250; e-mail pathnet@pathnet.org.

The statements and conclusions contained in this publication are those of the National Institute of Building Sciences and do not necessarily reflect the view of the Department of Housing and Urban Development. The Institute has made every effort to verify the accuracy and appropriateness of the publication's content. However, no guarantee of the accuracy or completeness of the information or acceptability for compliance with any industry standard or mandatory requirement of any code, law, or regulation is either offered or implied.

Foreword

An important factor in making the best use of our nation's housing stock is accurately assessing the condition, safety, usefulness, and rehabilitation potential of older residential buildings. The *Residential Rehabilitation Inspection Guide* provides step-by-step technical information for evaluating a residential building's site, exterior, interior, and structural, electrical, plumbing, and HVAC systems.

First published by the U.S. Department of Housing and Urban Development in 1984 as the *Guideline on Residential Building Systems Inspection*, the guideline has found widespread use and acceptance among architects, engineers, builders, realtors, and preservationists.

Now, for the Partnership for Advancing Technology in Housing (PATH) program, the guideline has been updated and expanded to include current assessment techniques and standards, information about additional building materials, and a broader coverage of hazardous substances and the effects of earthquakes, wind, and floods. HUD is pleased to reissue this important and time-tested publication, knowing that it will prove a valuable resource for preserving and reusing our nation's building stock.

Susan M. Wachter
Assistant Secretary for Policy Development and Research

Acknowledgments

The National Institute of Building Sciences (NIBS) produced the original edition of this guideline for the U.S. Department of Housing and Urban Development in 1984. It was written by William Brenner of Building Technology, Incorporated, with supplementary material and photographs provided by Richard Stephan, Ken Frank, and Gerard Diaz of the University Research Corporation. Technical reviewers were George Schoonover, Eugene Davidson, Joseph Wintz, Richard Ortega, Nick Gianopulos, Robert Santucci, James Wolf, and Thomas Fean.

This revised edition of the guideline was produced in 1999 by NIBS and updated and expanded by Thomas Ware and David Hattis of Building Technology, Incorporated. Technical reviewers were William Asdal, Neal FitzSimons, Wade Elrod, Hal Williamson, Paul Beers, John Bouman, Raymond Jones, Dan Kluckhuhn, Joe Sherman, William Freeborne, and Robert Kapsch. The graphic designer was Marcia Axtmann Smith. Selected illustrations are excerpted with permission from The Illustrated Home by Carson Dunlop & Associates (800-268-7070) and the material in Appendix C is used with the permission of the National Association of Homebuilders. William Brenner directed the project for NIBS and Nelson Carbonell was the HUD project manager.

Table of Contents

Foreword iii

Acknowledgments v

Introduction 1

1—Site 3
    1.1   Drainage 3
    1.2   Site Improvements 6
    1.3   Outbuildings 9
    1.4   Yards and Courts 9
    1.5   Flood Regions 9

2—Building Exterior 10
    2.1   Foundation Walls and Piers 10
    2.2   Exterior Wall Cladding 10
    2.3   Windows and Doors 12
    2.4   Decks, Porches, and Balconies 14
    2.5   Pitched Roof Coverings 16
    2.6   Low-Slope Roof Coverings 18
    2.7   Skylights 19
    2.8   Gutters and Downspouts 20
    2.9   Chimneys 22
    2.10 Parapets and Gables 23
    2.11 Lightning Protection 23

3—Building Interior 24
    3.1   Basement or Crawl Space 24
    3.2   Interior Spaces, General 26
    3.3   Bathrooms 29
    3.4   Kitchens 30
    3.5   Storage Spaces 31
    3.6   Stairs and Hallways 32
    3.7   Laundries and Utility Rooms 32
    3.8   Fireplaces and Flues 32
    3.9   Attics and Roof Truss and Joist Spaces 33
    3.10 Whole-Building Thermal Efficiency Tests 35
    3.11 Sound Transmission Control Between Dwelling Units 36
    3.12 Asbestos 36

	3.13 Lead	37
	3.14 Radon	38
	3.15 Tornado Safe Room	38

4—Structural System 39

4.1	Seismic Resistance	39
4.2	Wind Resistance	40
4.3	Masonry, General	41
4.4	Masonry Foundations and Piers	44
4.5	Above-Ground Masonry Walls	48
4.6	Chimneys	53
4.7	Wood Structural Components	54
4.8	Iron and Steel Structural Components	58
4.9	Concrete Structural Components	61

5—Electrical System 62

5.1	Service Entry	62
5.2	Main Panelboard (Service Equipment)	63
5.3	Branch Circuits	66

6—Plumbing System 69

6.1	Water Service Entry	69
6.2	Interior Water Distribution	71
6.3	Drain, Waste, and Vent Piping	73
6.4	Tank Water Heaters	76
6.5	Tankless Coil Water Heaters (Instantaneous Water Heaters)	78
6.6	Water Wells and Equipment	78
6.7	Septic Systems	80
6.8	Gas Supply in Seismic Regions	81

7—HVAC System 82

7.1	Thermostatic Controls	82
7.2	Fuel-Burning Units, General	84
7.3	Forced Warm Air Heating Systems	87
7.4	Forced Hot Water (Hydronic) Heating Systems	88
7.5	Steam Heating Systems	94
7.6	Electric Resistance Heating	96
7.7	Central Air Conditioning Systems	97
7.8	Central Gas-Absorption Cooling Systems	100
7.9	Heat Pumps	100

7.10 Evaporative Cooling Systems	101
7.11 Humidifiers	103
7.12 Unit (Window) Air Conditioners	103
7.13 Whole House and Attic Fans	103

Appendix A—The Effects of Fire on Structural Systems A-1

Appendix B—Wood-Inhabiting Organisms B-1

Appendix C— Life Expectancy of Housing Components C-1

Appendix D—References D-1

Appendix E—Inspection Record E-1

List of Figures

4.1 Assessing Structural Capacity	39
5.1 Assessing Electrical Service Capacity	63
6.1 Assessing Water Supply Capacity	70
6.2 Assessing DWV Capacity	74
6.3 Assessing Hot Water Heater Capacity	75
6.4 Assessing Well Capacity	79
6.5 Assessing Septic Capacity	80
7.1 Assessing Heating and Cooling Capacity	83

Residential Inspection Guideline

Introduction

The *Residential Inspection Guideline* is designed to help evaluate the rehabilitation potential of small residential buildings and structures. It may be used by contractors, builders, realtors, home inspectors, and others with a basic knowledge of building construction.

When used in conjunction with the local building code, the guideline can assist in identifying unsafe or hazardous conditions and uncovering functional deficiencies that should be corrected. It does not establish rehabilitation standards or address construction, operation, and maintenance costs.

Preparing for the Inspection

Before visiting the site, check with the local jurisdiction to determine:

- the site's **zoning, setback, height, and building coverage requirements, grandfathered uses and conditions, proffers, liens, and applicable fire regulations.**
- if the site is in a **seismic zone**.
- if the site is in a **hurricane** or **high tornado-risk region.**
- if the site is in a **flood plain** or other **flood-risk zone.**
- if there is any record of **hazards** in the soil or water on or near the site.

Conducting the On-Site Inspection

Once at the site, conduct a brief walk-through of the site and the building. Note the property's overall appearance and condition. If it appears to have been well maintained, it is far less likely to have serious problems. Note the building's style and period and try to determine when it was built. Next, examine the quality of the building's design and construction and that of its neighborhood. There is no substitute for good design and sound, durable construction. Finally, assess the building's functional layout. Does the building "work" or will it have to be significantly altered to make it usable and marketable?

Look for signs of dampness and water damage. Water is usually a building's biggest enemy and a dry building will not have problems with wood decay, subterranean termites, or rusted and corroded equipment.

After completing the initial walk-through, begin the formal inspection process:

- **Inspect the site, building exterior, and building interior** in accordance with Chapters 1, 2, and 3. Use the tests described in Chapters 2 and 3 when appropriate. Record pertinent information as needed.
- **Inspect the structural, electrical, plumbing, and HVAC systems** in accordance with Chapters 4, 5, 6, and 7. Use the tests described in each chapter as necessary. Record the size, capacity, and other relevant information about each system or component as needed.

While most inspections consist of observing, measuring, and testing building elements that are exposed to view, there are conditions that require the removal of some part of the building to observe, measure, or test otherwise concealed construction. Such intrusive inspections require some demolition and should be performed only with the permission of the owner and by experienced, qualified mechanics.

The building inspection forms in Appendix E may be copied for use during on-site inspections. Record general building data and site layouts, elevations, and floor plans first. This information will form the basis for later rehabilitation decisions. Then record the **size, capacity,** and **condition/needed repairs** information for each building component. This will highlight what needs to be repaired or replaced.

The inspection may be completed in one visit or over several visits, depending on the property's condition, the weather, problems of access, and the need for testing or expert help.

More Information

Appendix A provides information on assessing the effects of fire on wood, masonry, steel, and concrete structural systems. Appendix B can be used as an aid in the identification of wood-inhabiting molds, fungi, and insects. Appendix C lists the average life expectancies of common housing materials, components, and appliances. Appendix D provides ordering and Internet access information for the publications and standards referenced herein as well as a listing of applicable publications on building assessment, energy conservation, and historic preservation.

Use the *Secretary of the Interior's Standards for Rehabilitation* when dealing with historic properties. They are available full text online at http://www2.cr. nps.gov/tps.

When a property is rehabilitated for resale or when a contractor or builder is rehabilitating a property for its owner, consider using the *Residential Construction Performance Guidelines*. These were developed by the National Association of Home Builders' Remodelers Council, Single Family Small Volume Builders Committee.

When assessing the tornado risk at a site, consider using *Taking Shelter from the Storm: Building a Safe Room Inside Your House*, available from the Federal Emergency Management Agency (FEMA).

When assessing the flood risk at a site and before undertaking any applicable rehabilitation measures, consider using *Design Manual for Retrofitting Flood Prone Residential Structures*, available from the Federal Emergency Management Agency.

When inspecting a building located in a region of high seismic activity or in a hurricane region, additional information on vulnerability assessment and retrofit options can be found in *Is Your Home Protected from Earthquake Disaster?* and *Is Your Home Protected from Hurricane Disaster?* Both documents are available from the Institute for Business and Home Safety or can be viewed full text online at http://www.ibhs.org.

For those interested in working with local officials to make building codes more amenable to rehabilitation work, see the U.S. Department of Housing and Urban Development's *Nationally Applicable Recommended Rehabilitation Provisions*.

1
Site

Begin the rehabilitation inspection by thoroughly examining the property's drainage, site improvements, and outbuildings. Although their condition may have a profound impact on the total costs of the rehabilitation project, they are often overlooked or not fully considered in the initial building assessment. Tree removal, the replacement of sidewalks and driveways, and the repair of outbuildings can add substantially to rehabilitation expenses and may make the difference between a project that is economically feasible and one that is not.

Earthquake. Check the slope of the site. Buildings constructed on slopes of 20 degrees or more should be examined by a structural engineer in all seismic regions, including regions of low seismic activity. See Section 4.1, Seismic Resistance.

Wind. If the site is in a hurricane or high wind region, it should be examined for loose fences, tree limbs, landscaping materials such as gravel and small rocks, and other objects that could become windborne debris in a storm. See Section 4.2, Wind Resistance.

Floods. Five major flood-risk zones have been established to define where floods occur, and special flood resistance requirements have been created for each zone. Check with local authorities. See Section 1.5, Flood Regions.

Lead. Consider checking for the presence of lead in the soil, which can be a hazard to children playing outdoors and can be brought indoors on shoes. Lead in soil can come from different sources such as discarded lead-based paint, lead-based paint chips near foundations from when exterior walls were scraped and painted, leaded gasoline (now banned) on driveways where car repairs were made, leaded gasoline from car exhaust, and old trash sites where lead-bearing items were discarded. Check the site for evidence of any of these conditions and if found, consider having the soil tested for lead content.

Wildfires. In locations where wildfires can occur, some jurisdictions have requirements for hydrant locations and restrictions on the use of certain building materials as well as restrictions on plantings close to a building. Check with the local building official and the fire marshal for such requirements.

Building Expansion. If a rehabilitation project includes expanding a building or outbuilding, an assessment of the site for this work is critical. There is also a complementary need to examine zoning regulations to establish allowable coverage and setbacks. The use of available land may be restricted by coverage and setback requirements that define the areas on the site that can be used for new construction.

Site Restrictions. Homeowner association bylaws and deed covenants sometimes include requirements that can affect changes or additions to a building or outbuilding. These documents should be carefully examined to determine their impact.

Accessibility. When universal design is a part of a rehabilitation, consult the HUD publication *Residential Remodeling and Universal Design* for detailed information about parking, walks, and patios.

1.1
Drainage

Observe the drainage pattern of the entire property, as well as that of adjacent properties. The ground should slope away from all sides of the building. Downspouts, surface gutters, and drains should direct water away from the foundation. Check the planting beds adjacent to the foundations. Plantings are often mounded in a way that traps water and edging around planting beds acts like a dam to trap water. Most problems with moisture in basements are caused by poor site drainage.

The ground also should slope away from window wells, outside basement stairs, and other areaways. The bottom of each of these should be sloped to a drain. Each drain should have piping that connects it to a storm water drainage system, if there is one, or that drains to either a discharge at a lower grade or into a sump pit that collects and disperses water away from the building. Drains and piping should be open and clear of leaves, earth, and debris. A garden hose can be

Poor site drainage leads to a variety of problems, in this case a wet basement.

and catch basins should be such that if they became blocked and overflowed no significant damage will occur and that any resultant ice conditions will not pose a danger to pedestrians or vehicles. The design of surface drainage systems is based on the intensity and duration of rain storms and on allowable runoff. These conditions are usually regulated by the local building code, which can be used to check the adequacy of an existing surface drainage system.

In some locations, especially where slopes lack vegetation to slow water flow, it may be possible to reduce rehabilitation costs by diverting rainwater into a swale at or near the top of the slope and thereby reduce the amount of rainwater runoff handled by a surface drainage system. This swale, of course, must be within the property on which the building is located.

The ground beneath porches and other parts of a building that are supported on piers should be examined carefully. It should have no low areas and be sloped so that water will not collect there.

Water from the roof reaches the ground through gutters and downspouts or by flowing directly off roof edges. Because downspouts create concentrated sources of water in the landscape, where they discharge is important. Downspouts should not discharge where water will flow directly on or over a walk, drive, or stairs. The downspouts on a hillside building should discharge on the downhill side of the building. The force of

used to test water flow, although its discharge cannot approximate storm conditions.

Where a building is situated on a hillside, it is more difficult to slope the ground away from the building on all sides. On the high ground side of the building, the slope of the ground toward the building should be interrupted by a surface drainage system that collects and disposes of rainwater runoff. There are two general types of surface drainage systems: an open system consisting of a swale (often referred to as a ditch), sometimes with a culvert at its end to collect and channel water away, and a closed system consisting of gutters with catch basins. Combinations of the two are often used. The locations and layout of culverts, gutters, drains, water leaving a downspout is sometimes great enough to damage the adjacent ground, so some protection at grade such as a splash pan or a paved drainage chute is needed. In urban areas, it is better to drain downspouts to an underground storm water drainage system, if there is one, or underground to discharge at a lower grade away from buildings.

Water that flows directly off a roof lacking gutters and downspouts can cause damage below. Accordingly, some provision in the landscaping may be needed, such as a gravel bed or paved drainage way.

When a sump pump is used to keep a building interior dry, the discharge onto the site should be located so that the discharge drains away from the building and does not add to the subsurface water condition the sump pump is meant to control.

The site should be examined overall for the presence of springs, standing water, saturated or boggy ground, a high water table, and dry creeks or other seasonal drainage ways, all of which may affect surface drainage. It is especially important to inspect the ground at and around a septic system seepage bed, seepage pit, or absorption trenches. See Section 6.7.

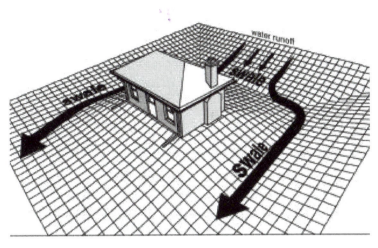

Where a building is situated on a hillside, swales can be used to direct surface water away from the foundation.

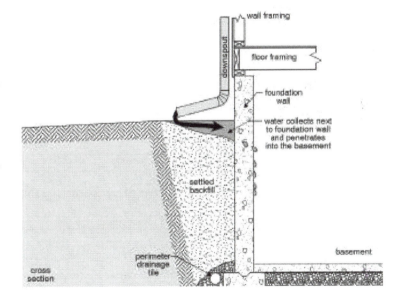

Settled backfill allows water to collect next to the foundation wall and penetrates into the basement.

1.2
Site Improvements

Well-maintained landscaping and other site improvements are important for the enjoyment, resale, or rental value of a property. Inspect the following:

- **Plantings.** Note the location and condition of all trees and shrubbery. Those that are overgrown may need pruning or trimming; in some cases they may be so overgrown that they will have to be removed. When trees or shrubbery exhibit disease or infestation, consult a qualified expert. Removing large trees may require special expertise and can be particularly costly.

    Check where overhanging branches may interfere with the chimney's draft, damage utility wires, or deposit leaves and twigs in roof gutters and drains.

    Trees and shrubbery that are very close to exterior walls or roofs can cause damage that is sometimes severe, and they can make it difficult to make inspections, do maintenance, and make repairs. Branches in these locations will need to be pruned back.

    Tree roots under paving and stairs can cause damage that is sometimes severe. Roots are usually exposed near the surface and will need to be cut back.

    Tree roots can heave foundations and may cause cracking by pushing against foundations from the outside. If tree roots are under a footing, cutting down the tree can lead to rotting of the roots and subsequent settling of the foundation.

    Observe the solar shading characteristics of all site plantings. Do they provide protection from the summer sun and allow the winter sun to warm the building? Large deciduous trees located to the south and west of a building can do both, and a special effort should be made to retain and protect such trees where they exist.

- **Fences.** Fences are usually installed to provide physical or visual privacy. Examine their plumbness and overall condition. Inspect wood fences for signs of rot or insect infestation and inspect metal fences for rust. Inspect all gates and their associated hardware for proper fit, operation, and clearance. Fences are often addressed in homeowner association bylaws and deed covenants. These should be checked and their requirements, if any, compared to existing conditions or used for the design of a new or replacement fence. Pay special attention to fence locations and property lines.

- **Lighting.** Examine outdoor lighting elements to determine their condition and functional safety. Turn site lighting on, preferably at night, to check its operation and to determine if the light is adequate for its purpose. Exposed wiring that is not UV- and moisture-resistant should be replaced. Underground wiring should be type UF. Fixtures, switches, and outlets should be properly covered and protected from moisture penetration.

- **Paved areas.** Inspect all walks, drives, and patios for their condition and to make sure paved areas immediately adjacent to a building are sloped away from building walls. Paving that is not sloped to drain water away from a building should be replaced. Inspect paving for cracks, broken sections, high areas, low areas that trap water, and tripping hazards.

    Paved areas that are made of concrete and are in poor condition may have to be replaced. Concrete cannot be repaired by resurfacing with a thin layer of more concrete. Concrete repairs in climates where freezing occurs should be no less than three inches thick. Where there is no freezing weather, repairs that are two inches thick may be used. Cracks in concrete should be cut open and sealed with a flexible sealant compound, which will extend its service life albeit not improve its appearance. Where there is a difference in elevation in a walk or drive that creates a tripping hazard, the higher portion of concrete may be ground down to the level of the lower portion, although the grinding will change the appearance of the concrete. Sunken areas of concrete paving result from failure of the subbase. For sidewalks it may be possible to lift up sections of the paving between construction joints, add to and compact the subbase to the proper elevation, and replace the paving sections.

Failed or sunken areas of asphalt drives and walks usually should be resurfaced or replaced. Sealing asphalt paving extends its life. Examine the paving to determine when sealing is needed. Check asphalt drives and walks for low areas that hold water and freeze in cold climates. Low areas in asphalt paving can be brought to level with an asphalt overlay.

Brick or stone patio paving should be set either on a concrete slab in a mortar bed with mortar joints or in a sand bed that is laid on earth or on a concrete slab. Mortar joints can be tuck pointed and loose bricks or stones can be reset in a new mortar bed. Pavers set in sand can be taken up easily, sand added or removed, and the pavers replaced.

When considering the repair or replacement of such site elements, pay particular attention to existing property lines and easements.

The maintenance, repair, and replacement of sidewalks, drive aprons, and curb cuts at the street may be the responsibility of the local jurisdiction. Check the property's deed or consult local authorities.

■ **Stairs.** Inspect the condition of exterior stairs and railings using the current building code as a guide. Every stair with more than three steps should have a handrail located 34 to 38 inches (865 to 965 mm) above the edges of the stair tread. Shake all railings vigorously to check their stability and inspect their fastenings. Stairs that are more than 30 inches (760 mm) above the adjacent grade and walks located more than 30 inches (760 mm) above the grade immediately below should have guards not less than 36 inches (915 mm) high and intermediate rails that will not allow the passage of a sphere 4 inches (100 mm) in diameter. Check wooden steps for proper support and strength and for rot and insect infestation. Inspect steel stairs for rust, strength, and attachment. Deteriorated stairs should be repaired or replaced. Stair treads should be as level as possible without holding water. It is preferable that stairs in walks on site that are accessible to the general public have at least three risers. Stair riser heights and tread depths should be, respectively, uniform.

■ **Retaining walls.** Inspect the construction and condition of retaining walls. Retaining walls more than two feet in height should be backed with drainage material, such as gravel. There should be drains at the bottom of the drainage material. The drains should discharge water either at the end of the wall or through pipes set in the wall itself. These drains and the drainage material behind the wall relieve the pressure of ground water on the wall. If possible, weep holes and related drains should be examined closely following a reasonably heavy rain to make sure they are working properly. If they are not discharging water, the drains should be cleaned out and observed again in the next rain. Failure to drain should be remedied by excavating behind the wall, replacing the drainage material and damaged drainage piping, and backfilling. In all but the driest climates, improper drainage of water from behind a retaining wall can cause the wall to fail.

Check for bowing (vertical bulges), sweeping (horizontal bulges), and cracking in retaining walls that can be caused by water pressure. Bulging can also be a result of inadequate strength to resist the load of the earth behind the wall. Bowing and sweeping failures may be correctable if found early enough and if the cause is poor drainage.

Check for other failures of retaining walls. Failure by overturning (leaning from the top) or sliding may be caused by inadequate wall strength. In addition, water behind a wall can create moist bearing, especially in clay soils, and contribute to sliding. Retaining walls also fail due to settling and heaving. The former occurs whenever filled earth below the wall compacts soon after the wall is built, or when wet earth caused by poor drainage dries out and soil consolidates at any time in a wall's service life. Poor drainage contributes to failure in cold climates by creating

The outward movement of the upper part of this retaining wall can be halted only by structural reinforcement. Simply patching the crack will not solve the problem.

heaving from frozen ground. Both overturning and sliding may be stabilized and sometimes corrected if the amount of movement is not extreme. Settling may be corrected on small, low walls of concrete or masonry, and heaving may be controlled by proper drainage. Significant failure of any kind usually requires rebuilding or replacing all or part of a wall. Failing retaining walls more than two feet in height should be inspected by a structural engineer.

■ **Buried oil tanks.** Buried ferrous metal oil tanks are common on older properties that have buildings or domestic water heated by oil. The presence of a buried oil tank usually can be determined by finding the fill pipe cover on the ground and the vent pipe that extends above ground to a height of at least four feet. Abandoned and very old buried ferrous metal oil tanks are an environmental hazard. If such a buried tank is located on the property, the soil around it should be tested by a qualified environmental engineer for the presence of oil seepage. If leaking has occurred, the tank and all contaminated soil around it must be removed. If leaking has not occurred, it may still be a potential problem. Even if a tank is empty, it still may have residual oil in the bottom that is a pollutant. Strong consideration should be given to removing the tank or filling it with an approved inert material after pumping out any old residual oil.

■ **Aerials.** On-site installations of aerial masts either from the ground or mounted to a tree or building should be assessed for structural stability, especially in high wind areas.

1.3 Outbuildings

Examine detached garages, storage sheds, and other outbuildings for their condition in the same way that the primary building is inspected. Check each outbuilding's water shedding capability and the adequacy of its foundations. On the interior, look for water staining on the roof or walls. Wood frame structures should be thoroughly inspected for rot and insect infestation. Check also that all doors function properly and that doors and windows provide adequate weather protection and security for the building. Make sure that small outbuildings have sufficient structural strength to sustain the applicable wind loads or seismic forces.

If the site is in a hurricane or high-wind region, check all outbuildings for their ability to resist a storm without coming apart and becoming windborne debris. Consider consulting an engineer.

1.4 Yards and Courts

In urban areas, two or more dwelling units may share a yard or court to provide light and ventilation to interior rooms. The adequacy of the light provided is a function of the dimensions of the yard or court, as well as the color of surrounding walls. Check these characteristics, as well as zoning and building and housing code requirements pertaining to light, ventilation, and privacy screening for yards and courts. Such requirements may affect the reuse of the property and their implications should be understood before the property is altered or purchased.

1.5 Flood Regions

The Federal Emergency Management Agency and the National Flood Insurance Program have established and defined five major flood-risk zones and created special flood resistance requirements for each.

Improperly designed grading and drainage may aggravate flood hazards to buildings and cause runoff, soil erosion, and sedimentation in the zones of lower flood risk, the Interflood Zone, and the Non-Regulated Flood Plain. In these locations, local agencies may regulate building elevations above street or sewer levels. In the next higher risk zones, the Special Flood Hazard Areas and the Non-Velocity Coastal Flood Areas (both Zone A), the elevation of the lowest floor and its structural members above the base flood elevation is required. In the zone of highest flood risk, the Coastal High Hazard Areas (Velocity Zone, Zone V), additional structural requirements apply.

Check with local authorities to determine if the site is in a flood-risk zone. If it is, check with local building officials. Higher standards than those set by national agencies have been adopted by many communities.

2
Building Exterior

After the site inspection has been completed, systematically inspect the building's exterior for its condition and weathertightness. Begin either at the foundation and work up or begin at the roof and work down. Examine the quality and condition of all exterior materials and look for patterns of damage or deterioration that should be further investigated during the interior inspection. Determine the building's architectural style and note what should be done to maintain or restore its integrity and character. See Chapter 4 for assessing structural components of the building.

In regions of medium to high seismic activity, buildings with irregular shapes (in either plan or elevation) may be especially vulnerable to earthquakes. Examine the building for such irregularities, and if present, consider consulting a structural engineer.

In hurricane regions, examine screen and jalousie enclosures, carports, awnings, canopies, porch roofs, and roof overhangs to determine their condition and the stability of their fastenings. Then examine the following four critical areas of the exterior to determine their condition and strength: roofs, windows, doors, and garage doors.

In locations where wildfires can occur, some jurisdictions have restrictions on the use of flammable exterior materials. Check with the local building official or the fire marshal, or both, for detailed information.

Additional information on the evaluation and treatment of historic building exteriors is presented in the *Secretary of the Interior's Standards for Rehabilitation*, available full text online at http://www2.cr.nps.gov/tps.

When universal design is a part of a rehabilitation, consult HUD publication *Residential Remodeling and Universal Design* for detailed information about entrances, doors, and decks.

2.1
Foundation Walls and Piers

Foundation walls and piers in small residential buildings are usually made of masonry and should be inspected for cracking, deterioration, moisture penetration, and structural adequacy. See Sections 4.3 and 4.4. Wood posts and columns and concrete foundations and piers should be inspected in accordance with Sections 4.7 and 4.9.

2.2
Exterior Wall Cladding

Exterior walls above the foundation may be covered with a variety of materials, including wood siding or its aluminum and vinyl substitutes, wood or asbestos cement shingles, plywood with and without a medium density (plastic) overlay, stucco, brick or stone masonry, and an exterior insulation and finish system. These materials are designed to serve as a weathertight, decorative skin and, in warm climates should be light in color to reduce heat absorption. Inspect exterior claddings as follows:

- **Exterior wood elements.** Inspect all painted surfaces for peeling, blistering, and checking. Paint-related problems may be due to vapor pressure beneath the paint, improper paint application, or excessive paint buildup. Corrective measures for these problems will vary from the installation of moisture vents to complete paint removal. Mildew stains on painted surfaces do not hurt the wood and may be cleaned with a mildew remover.

  All wood elements should be checked for fungal and insect infestation at exposed horizontal surfaces and exterior corner joints, as specified for wood structural components in Section 4.7.

  Check the distance between the bottom of wood elements and grade. In locations that have little or no snow, the distance should be no less than six inches. In locations with significant, lasting snow, the bottom of wood elements should be no less than six inches above the average snow depth.

- **Aluminum and vinyl siding.** Aluminum and vinyl siding may cover up decayed or insect-infested wood but otherwise are generally low maintenance materials. Check for loose, bent, cracked, or

Residential Inspection

A second layer of shingles has filled the former gap between roof and zsiding, causing the siding to deteriorate. Shingles are cupped and beginning to fail as well.

The stucco is beginning to erode on this structure due to a poor roof drainage detail. A longer scupper would solve this problem.

broken pieces. Inspect all caulked joints, particularly around window and door trim. Many communities require aluminum siding to be electrically grounded; check for such grounding.

- **Asbestos cement shingles.** Like aluminum and vinyl siding, asbestos cement shingles may cover decayed or insect-infested wood. Check for loose, cracked, or broken pieces and inspect around all window and door trim for signs of deterioration.
- **Stucco.** Check stucco for cracks, crumbling sections, and areas of water infiltration. Old and weathered cracks may be caused by the material's initial shrinkage or by earlier building settlement. New, sharp cracks may indicate movement behind the walls that should be investigated. Refer to Section 4.5 for problems with masonry walls. It is difficult to match the color of stucco repairs to the original stucco, so plan to repaint surrounding stucco work where sections are mended.

- **Brick or stone veneers.** Inspect veneers for cracking, mortar deterioration, and spalling. Refer to Sections 4.3 and 4.5 for the inspection of above-ground masonry walls.
- **Exterior insulation and finish systems (EIFS).** EIFS, also known as synthetic stucco, has been in widespread residential use since the early 1990s. It generally consists of the following product layers (moving outward): insulation board, mesh and base coat layer, finish coat, and sealant and flashing.

EIFS was originally designed as a nondraining water and moisture barrier system. A drainage-type EIFS that allows water and moisture to penetrate the surface and then drain away has been developed more recently. Most existing EIFS in residential applications is installed over wood framing and is of the nondraining type. Water leakage and consequent rotting of the wood framing have become serious problems in many installations, especially at wall openings such as windows and doors, where inadequate flashing details can allow water seepage into the wall interior.

Manufacturers of EIFS differ in their installation methods. Inspecting existing EIFS is difficult because it is a proprietary product and there are no standard construction details. Use a trained specialist to check for concealed water damage and rot.

Exterior walls of older buildings usually contain no thermal insulation. Examine behind the cladding when possible to determine the presence of insulation, if any, and assess the potential for insulating the exterior walls.

Where mildew and mold are evident on exterior cladding or where interior walls are damp, there is the possibility that condensation is occurring in the walls. Moisture problems generally occur in cold weather when outside temperatures and vapor pressures are low and there are a number of water vapor sources within the building. The presence of moisture may be a result of an improperly installed or failed vapor barrier, or no vapor barrier at all. If condensation is suspected, an analysis of the wall section(s) in question should be made. This analysis will provide the information necessary to make the needed repairs.

2.3
Windows and Doors

Windows and doors are the most complex elements of the building's exterior and should be inspected from the outside as follows:

- **Exterior doors** should be examined for their condition, overall operation and fit, and for the functionality of their hardware. Door types include hinged, single and double doors of wood, steel, aluminum, and plastic with and without glazing. Check wood and plastic doors that are not protected from the weather. These doors should be rated for exterior use.

In warm climates, jalousie doors may also be in use. Check these doors to make sure the louvers close tightly and in unison for weathertightness.

Some buildings use glass framed doors of fixed and operable panels that have wood, vinyl-covered wood, and aluminum frames. Check the track of these sliding doors for dents, breaks, and straightness. Check the glides of operable panels for wear and check the sealing of fixed panels for weathertightness. Note the degree of physical security offered by doors and their locksets and pay special attention to pairs of hinged and sliding doors.

Doors also should be inspected for the exterior condition of their frames and sills. Check doors that are not protected from the weather for the presence of essential flashing at the head.

Glazing on exterior doors should be examined as described in the following section on windows. The interior condition and hardware of exterior doors will be examined during the interior inspection.

In hurricane regions, check exterior doors, and especially double doors, for the presence of dead-bolt locks with a throw length of no less than one inch.

- **Windows** should be inspected for the exterior condition of their frames, sills and sashes, and for overall operation and

Residential Inspection

The glazing putty in this window is deteriorated in some locations. Repairs will be time consuming.

fit. The interior condition and hardware of windows will be examined during the interior inspection. There are eight types of windows and six types of frame material in general use in residential buildings. Frame materials are plastic, aluminum, steel, wood, plastic-clad wood, and metal-clad (steel or aluminum) wood. Window types are double hung, single hung, casement, horizontal sliding, projected out or awning, projected in, and fixed. In addition to these, there are jalousies: glass louvers on an aluminum or steel frame.

The glazing compound or putty around glass panels in older sashes should be examined especially carefully since this is often the most vulnerable part of the window and its repair is time consuming. Examine glazing tapes or strips around glass panels in steel or aluminum sashes for signs of deterioration such as hardened sealant or poor fit. Check metal sashes for weep holes that have been blocked by paint, sealant, or dirt. Weep holes are usually easy to clean. Check windows that are not protected from the weather for the presence of essential flashing at the head.

For windows close to the ground or easily accessible from flat roofs, note the degree of physical security provided by the windows and their locks.

In hurricane regions, check all windows and glass doors that are not protected by shutters to determine if they have been tested for impact resistance to windborne debris. If they have not been so tested, determine if plywood panels can be installed for their protection at the time of a hurricane warning.

■ **Weather stripping.** Window and door weather stripping is generally of three types: metal, foam plastic, or plastic stripping. Check each type for fit. Check metal for dents, bends, and straightness. Check foam plastic for resiliency and plastic stripping for brittleness and cracks. Make sure the weather stripping is securely held in place.

- **Shutters.** Window shutters are generally of two types: decorative and functional. Decorative shutters are fixed to the exterior wall on either side of a window. Check the shutter's condition and its mounting to the wall. Functional shutters are operable and can be used to close off a window. Assess the adequacy of these shutters for their purpose: privacy, light control, security, or protection against bad weather. Check their operation and observe their condition and fit.

  Shutters close to the ground can be examined from the ground. Shutters out of reach from the ground should be examined during the interior inspection when windows are examined.

  In hurricane regions, check shutters to see if the shutter manufacturer has certified them for hurricane use. If they provide protection to windows and glass doors, determine if they have been tested for impact resistance to windborne debris.

- **Awnings.** Windows and glazed exterior doors sometimes have awnings over them, usually for sun control, but sometimes for decoration or protection from the weather. Awnings are usually made of metal, plastic, or fabric on a metal or plastic frame. Some are fixed in place, while others are operable and can be folded up against the exterior wall. Check the condition of awnings. Assess the adequacy of the attachment to the exterior wall. Fold up and unfold operable awnings and note the ease of operation. If an awning is used for sun control, assess its effectiveness and its effect on energy conservation.

- **Storm windows and doors** should be examined for operation, weathertightness, overall condition, and fit. Check the condition of screen and glass inserts; if they are in storage, locate, count, and inspect them. Check also to determine if the weep holes have been blocked by paint, sealant, dirt, or other substances. Opening weep holes is usually easy to do.

- **Garage doors** should be examined for operation, weathertightness, overall condition, and fit. Doors without motors should be manually opened and closed. Doors with motors should be operated using each of the operators on the system (key lock switch or combination lock key pad where control must be accessible on the exterior, remote electrical switch, radio signal switch, or photoelectric control switch). Check the operation for smoothness, quietness, time of operation, and safety. Check for the presence and proper operation of the door safety reversing device. Observe exposed parts of the installation for loose connections, rust, and bent or damaged pieces.

  Garage doors are made of wood, hardboard on a wood frame, steel, glass fiber on a steel frame, glass fiber, and aluminum. All come with glazed panes in a wide variety of styles. Check wood and hardboard for rot and water damage, check hardboard for cracking and splitting, check steel for rust, check glass fiber for ultraviolet light deterioration, and check aluminum for dents.

  In hurricane regions, examine garage doors, especially single doors on two-car garages, to determine if the assembly (door and track) has been tested for hurricane wind loads or has been reinforced.

- **Safety Glazing.** Glazed entrance doors including storm doors, sliding glass patio doors, and glazing immediately adjacent to these doors, but excluding jalousie doors, should be fully tempered, wire, or laminated glass or an approved plastic material. In addition, glazing adjacent to any surface normally used for walking must be safety glazing. Safety glazing is a building code requirement that applies to both new and replacement glazing.

2.4 Decks, Porches, and Balconies

Decks, porches, and balconies are exposed to the elements to a greater extent than most other parts of a building and are therefore more susceptible to deterioration. Inspect for the following:

- **Condition.** Examine all porch, deck, and balcony supports for signs of loose or deteriorated components. See Section 4.7 for the inspection of wood structural components. Masonry or concrete piers should be plumb and stable; check them in accordance with Section 4.4. Make sure that structural connections to the building are secure and protected against corrosion or decay.

Examine porch floors for signs of deflection and deterioration. Where the porch floor or deck is close to the level of the interior floor, look for signs of water infiltration at the door sill and check for a positive pitch of the porch floor or deck away from the exterior wall.

- Exterior railings and stairs. Inspect the condition of all exterior stairs and railings. Every stair with more than three steps should have a handrail located 34 to 38 inches (865 to 965 mm) above the edges of the stair tread. Shake all railings vigorously to check their stability, and inspect their fastenings. Most codes for new construction require that porches, balconies, and decks located more than 30 inches (760 mm) above the ground have guards not less than 36 inches (915 mm) high and intermediate rails that will not allow the passage of a sphere 4 inches (100 mm) in diameter. Check wooden steps for proper support and strength and for rot and insect infestation. Inspect steel stairs for rust, strength, and attachment. Deteriorated stairs should be repaired or replaced.

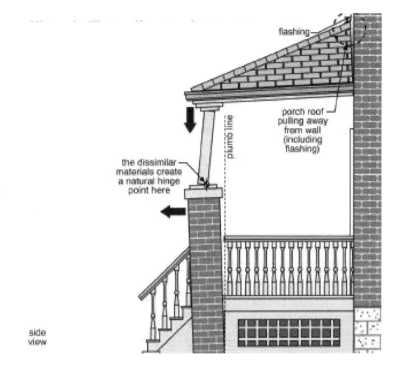

The joint between the two parts of this support creates a hinge that can affect the roof structure.

A rotted corner post on a screened porch. In this case, the rotted section of the post and a small section of the floor beneath it were removed and replaced with sound wood.

Stair treads should be as level as possible without holding water. Stair riser heights and tread depths should be, respectively, uniform.

2.5 Pitched Roof Coverings

Pitched or steep sloped roofs are best inspected when direct access is gained to all their surfaces. Use binoculars to inspect roofs that are inaccessible or that cannot be walked on. Look for deteriorated or loose flashing, signs of damage to the roof covering, and valleys and gutters clogged with debris. Carefully examine exterior walls and trim for deterioration beneath the eaves of pitched roofs that have no overhang or gutters. There are four categories of pitched roof covering materials and their condition should be checked as follows:

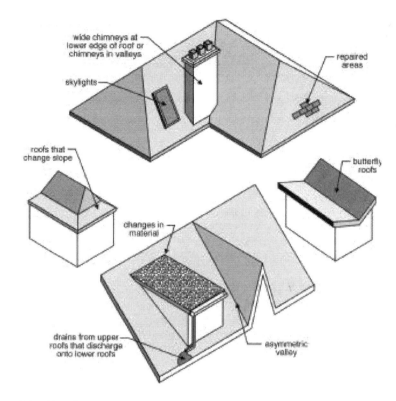

Vulnerable roof areas

- **Asphalt shingles.** Asphalt or "composition" shingles have a service life of about 20 years for the first layer and about 15 years for a second layer added over the first layer, depending on their weight, quality, and exposure. When they begin to lose their granular covering and start to curl they should be replaced. No more than two layers of asphalt shingles should normally be in place at any one time. If a second layer of asphalt shingles has been applied, check to see if all the flashing materials (galvanized steel, aluminum, rubber) in the first layer were removed and replaced with new flashing at the second layer.

  Check the roof slope. A slope of 4 in 12 or steeper is referred to as normal. A slope of between 3 in 12 and 4 in 12 is referred to as low. No asphalt shingle roof should be less steep than 3 in 12. If the roof has a normal slope, check the underlayment if possible. It should be at least a single layer of 15-pound (6.8 kg) asphalt saturated felt. Low-slope roofs should have at least two such felt layers. If ice dam flashing at overhanging eaves is needed (see Section 2.8) or present, make sure it extends three feet beyond the plane of the interior face of the exterior wall below for a low-slope roof and two feet for a normal-slope roof.

- **Wood shingles or shakes.** This type of covering has a normal life expectancy of 25 to 30 years in climates that are not excessively hot and humid, but durability varies according to wood species, thickness, the slope of the roof, whether shingles are made of heartwood, and whether they have been periodically treated with preservative. Shakes are hand-

This slate roof should be carefully investigated since it has a makeshift repair. Other problems include the chimney, which is too low, and the vent pipe, which is too narrow.

split on at least one face and either tapered or straight. Shingles are sawn and tapered. Check the roof slope. The minimum slope for wood shingles is 3 in 12 and the minimum slope for shakes is 4 in 12. As wood shingles and shakes age, they dry, crack, and curl. In damp locations they rot. Replace them when more than one-third show signs of deterioration. These materials are easily broken. They should not be walked on during the inspection. If the roof is historic or relatively complex, consult a wood roofing specialist.

■ **Metal roofing.** Metal can last 50 years or more if properly painted or otherwise maintained. Metal roofs may be made of galvanized iron or steel, aluminum, copper, or lead; each material has its own unique wearing characteristics. Inspect metal roofs for signs of rusting or pitting, corrosion due to galvanic action, and loose, open, or leaking seams and joints. The slope of metal roofing can be from one-half inch per foot (1:24) to very steep. The types of metal, seams, and slope determine the construction details. There are three basic seam types—batten, standing, and flat—as well as flat and formed metal panels. Snow guards are needed on steeper slopes and in locations with heavy, long-lasting snow, bracket and pipe snow guards also may be necessary. Low-slope metal roofs that are coated with tarlike material are probably patched or have pin holes and cannot be counted on to be leak-free. If the roof is historic or relatively complex, consult a metal roof specialist.

- **Slate, clay tile, and asbestos cement shingles.** These roof coverings are extremely durable and, if of high quality and properly maintained, may last the life of the structure. Check the roof slope. The minimum slope for roofs of these materials is 4 in 12. Slate shingles should be secured by copper nails except in the very driest of climates; look at the underside of the roof sheathing in the attic or check the nails on broken shingles. Nail heads should be covered with sealant. Nails for tile roofs should be non-corroding. All of these roof coverings are brittle materials and easily broken, and should not be walked on during the inspection. Use binoculars to look for missing, broken, or slipping pieces. Slate is particularly susceptible to breakage by ice or ice dams in the winter, and should therefore be especially well drained. Snow guards are needed on steeper slopes, and in locations with heavy, long-lasting snow, snow guards also may be necessary. Moss will sometimes grow on asbestos cement shingles; it should be removed with a cleaner to prevent capillary water leaks. Slate, clay tile, and asbestos shingles should be repaired or replaced by a qualified roofer.

Examine the underside of the roof later during the interior inspection.

2.6 Low-Slope Roof Coverings

A roof that is nearly level or slightly pitched is called a low-slope roof. No roof should be

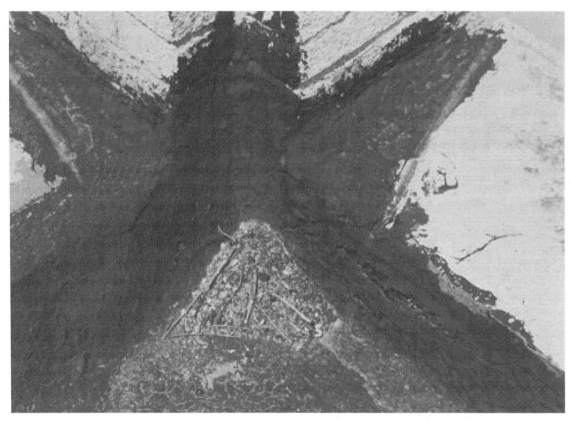

The built-up roof and flashings in this photograph are in poor condition. Patching may work temporarily, but the roof and flashings should be replaced.

dead level flat; it must have at least a slight slope to drain. Problems in low-slope roofs are common and more difficult to diagnose than pitched roof problems because the path of water leakage through flat roofs is often quite hard to trace. Look for signs of ponded water due to either improper drainage or sagging of the roof deck. If the cause is a sagging deck, it should be structurally corrected before it worsens. Low-slope roofs are expensive to repair, so extra care should be taken in their examination.

Inspect the flashing and joints around all roof penetrations, including drains, soil stacks, chimneys, skylights, hatchways, antenna mountings, and other roof-mounted elements. Note if metal flashings need painting or reanchoring and if asphaltic or rubber flashings are brittle or cracked. Check parapet wall caps and flashing for signs of damage due to wall movement.

Examine all portions of the roof covering. Look for signs of previous repairs that may indicate trouble spots. There are four categories of low-slope roof covering materials and they should be inspected as follows:

- **Built-up roofing.** Built-up roofs are composed of several layers of roofing felt lapped and cemented together with bituminous material and protected by a thin layer of gravel or crushed stone. Built-up roofs vary greatly in life span, but those used in residential buildings usually last about 20 years, depending on their quality, exposure, number of plies, and the adequacy of their drainage. Because built-up roofs are composed of several layers, they can contain moisture in the form of water or water vapor between layers. Moisture not only accelerates deterioration, it can also leak into a building. Look for cracking, blistering, alligatoring, and wrinkling, all of which may indicate the need for roof replacement or repair. Consult an experienced roofer for a further evaluation if you are in doubt.

  Test: An infrared or nuclear scanner can be used to detect areas of moisture in built-up roofs. Once located, these areas can be more thoroughly checked with a moisture meter or a nuclear meter. Such tests must be performed by a trained roofing inspector and are normally used to determine areas that need replacement on very large roofs.

- **Single-ply membrane roofing.** A single-ply membrane roof consists of plastic, modified bitumen, and synthetic rubber sheeting that is laid over the roof deck, usually in a single ply and often with a top coating to protect it from ultraviolet light degradation. Single-ply roofs are installed in three basic ways: fully adhered, mechanically attached, and loose laid with ballast. If properly installed and properly maintained, a single-ply roof should last 20 years. Roof penetrations and seams are the most vulnerable parts of single-ply membrane roofing and should be carefully checked. The material is also susceptible to ultraviolet light deterioration. A protective coating can be used to protect it, but the coating should be reapplied periodically. Check carefully for surface degradation on an unprotected roof and fading of the coating on a protected roof. Check also for signs of water ponding and poor drainage.

- **Roll roofing.** Roll roofing consists of an asphalt-saturated, granule-covered roofing felt that is laid over the roof deck. It can only provide single- or two-ply coverage. Inspect roll roofing for cracking, blistering, surface erosion, and torn sections. Seams are the most vulnerable part of roll roofing, and should be carefully checked for separation and lifting. Also check for signs of water ponding and poor drainage.

- **Metal roofing.** See Section 2.5.

The underside of the low-slope roof should be examined during the interior inspection. If it is inaccessible, look for signs of water leakage on interior ceilings and walls.

2.7 Skylights

From the exterior, check all skylights for cracked or broken glazing material, adequate flashing, and rusted or decayed frames. Skylights will be checked again during the interior inspection. Leaking skylights are common. Replacement skylights must comply with the building code.

The gutters on this low-slope roof are deteriorating largely because of the accumulated detritus that they hold. They should be inspected and cleaned periodically.

2.8
Gutters and Downspouts

Buildings with pitched roofs can have a variety of drainage systems. With a sufficient overhang, water can drain directly to the ground without being intercepted at the roof edge. See Section 1.1. Usually, pitched roofs end in gutters that are drained by downspouts.

Low-slope roof drainage is accomplished in one of three ways: without gutters or downspouts, with gutters and downspouts, or by downspouts that go down through a building's interior. Drainage without gutters and downspouts can damage the exterior wall with overflow. If the roof has no gutters and downspouts or interior downspouts, carefully examine the exterior walls for signs of water damage.

Gutter and downspout materials are usually galvanized steel, aluminum, copper, or plastic.

- **Gutters** should have a minimum ratio of gutter depth to width of 3 to 4; the front edge should be one-half inch (13 mm) lower than the back edge; and four inches is considered the minimum width except on the roofs of canopies and small porches. Make certain all gutters are clean and slope uniformly, without low areas, to downspouts. If there is a screen or similar device to prevent anything but water from flowing into the gutter, check its condition, fit, and position, to be sure water really can enter the gutter. Check gutters without screens or similar devices to be sure that basket strainers are installed at each downspout. Check the physical and functional

Eave protection against ice dams

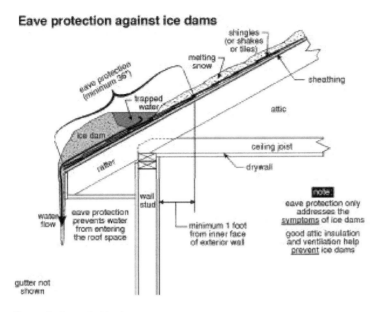

Eave protection against ice dams

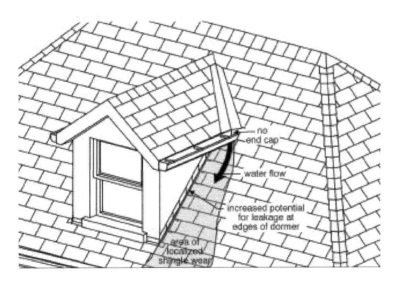

Dormer gutters improperly discharging onto the roof

condition of all gutters. Joints should be soldered or sealed with mastic. Also examine the placement of gutters: the steeper the roof pitch, the lower the gutter placement. On roofs with lower slopes make sure gutters are placed close to the roof's surface. Hangers should be placed no more than three feet apart. Where ice and snow are long lasting, hangers should be placed no more than 18 inches (460 mm) apart. Wherever a gutter is exposed, check the strength of its fastening to the roof fascia or building exterior. Rusted fasteners and missing hangers should be replaced.

■ **Ice dams** can form on pitched roof overhangs in cold climates subject to prolonged periods of freezing weather, especially those climates with a daily average January temperature of 30 °F (-1 °C) or less. Heat loss through the roof and heat from the sun (even in freezing temperatures) can cause snow on a roof to melt. As water runs down the roof onto the overhang, it freezes and forms an ice dam just above the gutter. The ice dam traps water from melting snow and forces it back under the shingles and into the building's interior. Check the edge of the roof overhang for evidence of ice dams and observe the eaves and soffit for evidence of deterioration and water damage. Check gutters and the immediately adjacent roofing for the presence of electrical de-icing cables, which may be evidence of an ice dam problem. When the interior inspection is made, check the inside of exterior walls and adjacent ceilings for signs of water damage. If the house has an attic, check the underside of the roof deck at exterior walls for signs of water damage.

- **Downspouts** should be checked for size. Seven square inches is generally the minimum except for small roofs or canopies. Check downspout attachments; there should be attachments or straps at the top, at the bottom, and at each intermediate joint. Check straps for rust, deformation, and failed or loose fasteners. Check the capacity of the drainage system. At least one downspout is usually needed for each 40 feet (12 m) of gutter. For roofs with gutters, make sure that downspouts are clear and that they discharge so water will drain away from the foundation. See Section 1.1. For low-slope roofs without gutters, interior downspouts cannot be examined from the roof, but check that basket strainers are in place. During the interior inspection, examine areas through which interior downspouts pass for signs of water damage.

On buildings with multiple roofs, one roof sometimes drains to another roof. Where that happens, water should not be discharged directly onto roofing material. Check to be sure that water is always directed to a gutter and that higher gutters discharge to lower gutters through downspouts.

Occasionally, wooden gutters and downspouts are used, usually in older or historic residences. They may be built into roof eaves and concealed by roof fascias. Wooden gutters are especially susceptible to rot and deterioration and should be carefully checked.

Pitched roofs in older buildings may end at a parapet wall with a built-in gutter integrated with the roof flashing. Here, drainage is accomplished by a scupper (a metal-lined opening through the parapet wall that discharges into a leader head box that in turn discharges to a downspout). Check the leader head box to be sure it has a strainer. Check the scupper for deterioration and open seams and check all metal roof flashings, scuppers, leader head boxes, and downspouts to make certain they are made of similar metals.

2.9 Chimneys

Chimneys should project at least two feet above the highest part of a pitched roof and anything else that is within 10 feet (3 m). A chimney should project at least three feet from its penetration from the roof (required minimum heights may vary slightly). Check the local building code. If the chimney is not readily accessible, examine what you can with binoculars from the highest vantage point you can find.

Flues should not be smaller in size than the discharge of the appliance they serve. The minimum flue area for a chimney connected to a fireplace is normally 50 square inches (320 cm^2) for round linings, 64 square inches (410 cm^2) for rectangular linings, and 100 square inches (650 cm^2) for an unlined chimney. Be extremely cautious about unlined chimneys; check the local building code. Flues should extend a minimum of four inches above the top of a masonry chimney. The height between adjacent flues in a multiple flue chimney without a hood should vary approximately four inches to avoid downdrafts. The same is true of a chimney with a hood unless a withe of masonry completely separates every flue.

Masonry chimneys without hoods should have stone or reinforced concrete caps at the top. Cement washes with or without reinforcing mesh are also used, but they are the least durable. Some masonry chimneys have hoods over the flues. Hoods on masonry chimneys consist of stone or reinforced concrete caps supported on short masonry columns at the perimeter of chimney tops, or sheet metal caps supported on short sheet metal columns. The height of a hood above the top of the highest flue should be at least 25 percent greater than the narrowest dimension of the flue.

Check the condition of chimney tops and hoods. If a cement wash is not properly sloped or is extensively cracked, spalled, or displays rust stains, it should be replaced. Reinforced concrete caps and stone caps with minor shallow spalling and cracking should be repaired. Those with extensive spalling or cracking should be replaced. Sheet metal hood caps with minor rust or corrosion should be repaired, but if rust or corrosion is extensive, replacement is needed.

Metal spark screens are sometimes used on wood and coal-burning fireplace chimneys. Check the condition and fit of spark screens. Dirty or clogged screens adversely affect draft and should be cleaned.

Where a masonry chimney is located on the side of a pitched roof, a cricket is needed on the higher side to divert water around the chimney. Check the cricket to be sure that its seams are watertight, that it is properly flashed into the chimney and roofing, and that it extends the full width of the chimney.

In seismic zones, check the bracing of masonry chimneys from the top of the firebox to the cap, and particularly the portion projecting above the roof. Consider consulting a structural engineer to determine the need for additional bracing or strengthening.

If the chimney is prefabricated metal encased in an exterior chase of siding, check the chase top to be sure it is properly interlocked with the metal chimney's counterflashing so that the assembly is watertight. Also check the chase top for slope: water should drain off the enclosure. Check for the presence of a terminal metal rain cap and make certain the flue terminates not less than two inches and not more than eight inches above the enclosure top.

If the chimney is prefabricated metal and not encased, check the adjustable flashing at the roof to be sure it is tightly sealed to the chimney, preferably with counterflashing, and check for the presence of a stack cap.

2.10
Parapets and Gables

In seismic zones, check the bracing of masonry parapets and gables. Consider consulting a structural engineer to determine the need for additional bracing or strengthening.

2.11
Lightning Protection

Lightning is a problem in some locations. Check the local building code. Lighting protection may be required to prevent powerline surge damage to electrical service, telephone service, or radio and television leads; to protect tall trees close to buildings; or to protect an entire building.

A lightning protection system is an interconnected aggregation of lightning rods, bonding connections, arresters, splicers, and other devices that are installed on a building or tree to safely conduct lightning to the ground. Lightning protection components and systems are identified by Underwriters Laboratories in three classes. Class I includes ordinary buildings (including residences) under 75 feet (22 m) in height. A Class I lightning protection system consists of lightning rods located on the roof and on projections, such as chimneys; main conductors that tie the lightning rods together and connect them with a grounding system; bonds to metal roof structures and equipment; arresters to prevent powerline surge damage; and ground terminals, usually rods or plates driven or buried in the earth. Lightning protection systems should be examined by a certified technician.

Using Appendix D, consult the following sources for technical information about lightning protection systems:

- Lightning Protection Institute, *Installation Code*, LPIB175.
- National Fire Protection Association, NFPA 780, *Standard for the Installation of Lightning Protection Systems*.
- Underwriters Laboratories, Inc., UL 96, *Lightning Protection Components* and UL 96A, *Installation Requirements for Lightning Protection Systems*.

3
Building Interior

Following the inspection of the site and the building's exterior, move indoors and systematically inspect all interior spaces, including basement or crawl space, finished rooms, halls and stairways, storage spaces, and attic. Begin either at the lowest level and work up or at the attic and work down. Examine the overall quality and condition of the building's construction and finish materials. If the interior has unique woodwork or other stylistic features, consider how these may be incorporated to best advantage in the building's reuse. Look for patterns of water damage or material deterioration that indicate underlying problems in the structural, electrical, plumbing, or HVAC systems. These systems will be inspected separately after the interior inspection has been completed.

When universal design is a part of a rehabilitation project, consult HUD publication *Residential Remodeling and Universal Design* for detailed information about doors, kitchens, bathrooms, laundry areas, closets, stairs, windows, and floor surfaces.

3.1
Basement or Crawl Space

The basement or crawl space is often the most revealing area in the building and usually provides a general picture of how the building works. In most cases, the structure is exposed

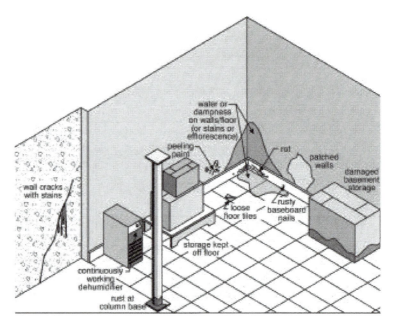

Clues to water problems in basements

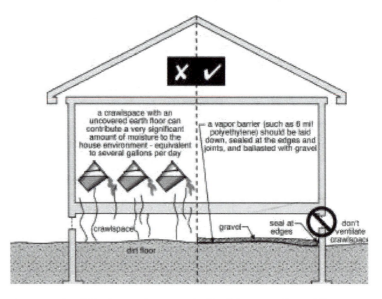

An uncovered earth floor in a crawl space can significantly increase moisture within the building.

Residential Inspection

Termite infestation is most common in basements and crawl spaces, particularly near foundation walls. Probe all suspect areas thoroughly.

overhead, as are the HVAC distribution system, plumbing supply and DWV lines, and the electrical branch circuit wiring.

- **Moisture.** One of the most common problems in small residential structures is a wet basement. Examine walls and floors for signs of water penetration such as dampness, water stains, peeling paint, efflorescence, and rust on exposed metal parts. In finished basements, look for rotted or warped wood paneling and doors, loose floor tiles, and mildew stains.

  Determine the source of any moisture that may be present. It may come through the walls or cracks in the floor, or from backed-up floor drains, leaky plumbing lines, or a clogged air conditioner condensate line. If moisture appears to be coming through the walls, re-examine the roof drainage system and grading around the exterior of the building (the problem could be as simple as a clogged gutter). Recheck the sump pump, if there is one, to be sure the discharge is not draining back into the basement. Look for unprotected or poorly drained window wells, leaking exterior faucets, and signs of leakage in the water supply line near the building. See Section 6.2 for water distribution system problems. If foundation walls are cracked, examine them in accordance with Section 4.4.

  Check the elevation of an earthen floor in a crawl space If the water table on the site is high or the drainage outside the building is poor, the crawl space floor should not be below the elevation of the exterior grade.

  If the basement or crawl space is merely damp or humid, the cause simply may be lack of adequate ventilation, particularly if the crawl space has an earthen floor.

  Check the ventilation. By measurement and calculation, compare the free area of vents with the plan area of the crawl space. The free vent area to crawl space area ratio should be 1 to 150 in a crawl space with an earthen floor and 1 to 1,500 in a crawl space with a vapor barrier of one perm or less over the earthen floor. If the calculated ratio is less, consider adding ventilation, particularly in hot and humid climates, and especially if moisture is present.

  Check the location of the vents through the foundation or exterior wall. There should be one vent near every corner of the crawl space to promote complete air movement. Check vents for screens. They should have corrosion resistant mesh in good condition with maximum 1/8-inch (3.2 mm) openings. If the ventilation appears to be inadequate and additional vents cannot be cut in the foundation or exterior wall economically, consider adding a vapor barrier and mechanical ventilation.

- **Fungal and insect infestation.** Look for signs of fungal growth on wood, particularly in unventilated crawl spaces.

Inspect all foundation walls, piers, columns, joists, beams, and sill plates for signs of termites and other wood inhabiting insects in accordance with Section 4.7. Also see Appendix B, Wood Inhabiting Organisms.

- **Thermal insulation.** Examine the amount and type of insulating material, if any, above unheated basements and crawl spaces. Determine the amount of insulation required for the space and whether additional insulation can or should be added. Check for adequate vapor barriers.

- **Structural, electrical, plumbing, and HVAC systems.** Understand enough about the layout of each system to make an informed inspection of the remainder of the building's interior. A more complete assessment of these systems will be performed later.

❏ *Note the type of structural system* (wood frame, masonry bearing wall, etc.). Locate main support columns and posts, major beams, and bearing walls.

❏ *Find the main electrical panel box,* if it is in the basement, and note how the branch circuits are generally distributed. Note also the type of wiring that is used.

❏ *Trace the path of the main water supply* line and check the composition of all piping materials.

❏ *Observe the general location of the heating/cooling unit,* if it is in the basement, and the general layout of the HVAC distribution system.

❏ *Locate the access to the crawl space,* check that it is large enough for a person to enter, observe the interior of the crawl space, and if mechanical equipment is located inside, check that access is large enough for any required maintenance.

3.2
Interior Spaces, General

This section deals with inspection procedures that are common to all interior spaces, including finished attics and basements. Examine the following elements and conditions of interior spaces:

- **Walls and ceilings.** Check the general condition of all surfaces, ignoring cosmetic imperfections. Look for cracks and peeling paint or wallpaper. Note signs of exterior water penetration or interior leakage. Whenever possible, probe behind wallpaper, paneling, ceiling tiles, and other coverings for problems that may have been concealed but not corrected.

Look for sags and bulges in old plaster work. Gently tap and push on the plaster; if an area sounds hollow or feels flexible, it is a good indication that the plaster has separated from its backing. If such areas

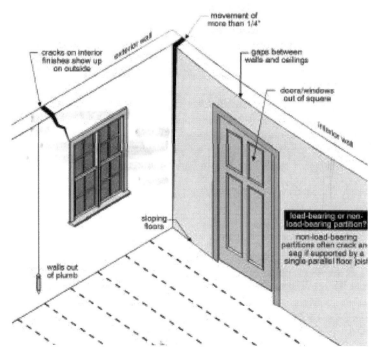

Interior clues to structural problems

are found, it may be best to replaster or overlay the wall or ceiling with wallboard.

Wall and ceiling cracks are usually caused by building settlement, deflection, warping of wood structural elements, or small seasonal movements of building components due to temperature and humidity variations. Seasonal movements will make some cracks regularly open and close; these may be filled with a flexible, paintable sealant, but otherwise cannot be effectively repaired. Cracks due to settlement, deflection, or warping can be repaired if movement has stopped, as is often the case.

Large wall and ceiling cracks may indicate structural problems. See Sections 4.3 through 4.5 for cracks associated with masonry wall problems and Section 4.7 for cracks associated with structural wood framing problems.

Inspect drywall-covered walls and ceilings by checking for nail popping, joint cracks, and other signs of deterioration or failure, such as rust stains at fasteners and corner beads.

Examine paneled walls by pushing or tapping on the paneling to determine if it is securely attached. Look for delamination of veneers. If the paneling is obviously not original, try to look behind it to see what problems may be covered up.

Lift suspended ceiling panels and observe above them. Check the condition of the original ceiling, if any. Tiled ceilings should be examined similarly. On top floors, inspect for ceiling penetrations that may form thermal bypasses to the unconditioned spaces above.

■ **Exterior walls.** In buildings built after 1960, try to determine if the exterior walls are insulated and contain a vapor barrier. Vapor barriers should be placed on the interior side of the insulation in cold climates and on the exterior side of the insulation in warm, moist climates.

■ Floors. Examine the floor's finish or covering. Inspect hardwood floors to determine if they will need cleaning or sanding. If sanding is required, be sure to check (by removing a floor register or piece of baseboard trim) how much the floor thickness has been reduced by previous sandings. Too much sanding will expose floor nails and, if present, tongue-and-groove joints.

Inspect resilient floors and carpeting for their overall condition and quality. If they are to be replaced, check that their floor underlayment is sound.

If the floor feels springy, sagging, or unstable, inspect it in accordance with Section 4.7.

Check the heating source in every room. This particular heater, when tested, was operable and safe.

- **Interior doors.** Inspect the condition of doors and door frames including the interior of entrance doors and storm doors. Check hardware for finish, wear, and proper functioning. Binding doors or out-of-square frames may indicate building settlement. See Section 4.4.

- **Windows.** Inspect window sash and frames for damage and deterioration. Operate each window, including storm windows and screens, to determine smoothness, fit, and apparent weathertightness. Pay particular attention to casement windows. When open they are easily damaged by wind and hinge damage may keep them from closing properly. Also carefully check casement operating hardware to be sure it operates smoothly and easily. Note the type and condition of glass in each window and assess its effect on energy use. If possible, determine if the window has a thermal break frame. Check for the presence and adequacy of security hardware. Examine the functioning of sash cords and weights in older double hung windows. Open windows above the ground floor (or others not fully inspected from the outside) and check their exterior surfaces, frames, sills, awnings, and shutters, if any.

Test: Air infiltration through windows and doors can be checked by the test method described in ASTM E783, *Standard Test Method for Field Measurement of Air Leakage Through Installed Windows and Doors.* The test should be performed by an experienced technician.

Test: Water penetration through windows and doors can be checked by the test method described in ASTM E1105, *Standard Test Method for Field Determination of Water Penetration of Installed Exterior Windows, Curtain Walls, and Doors by Uniform or Cyclic Static Air Pressure Difference.* The test should be performed by an experienced technician.

Consider window-related code requirements for natural light, ventilation, and egress capability. Most codes require the following:

❑ *Natural light.* Habitable rooms should be provided with natural light by means of exterior glazed openings. The area required is a percentage of the floor area, usually eight percent.

❑ *Ventilation.* Habitable rooms should be provided with operable windows. Their required opening size is a percentage of the floor area, usually four percent. A mechanical ventilation system can be provided in lieu of this requirement.

❑ *Egress.* Every sleeping room and habitable basement room should have at least one operable window or exterior door for emergency egress or rescue. Egress windows should have a minimum net clear opening of 5.7 square feet (0.53 m2), with a clear height of at least 24 inches (610 mm), a clear width of at least 20 inches (510 mm), and a sill height not more than 44 inches (1120 mm) above the floor. Emergency egress or rescue windows and doors should not have bars or grilles unless they are releasable from inside without a key, tool, or special knowledge.

- **Closets.** Inspect all closets for condition and usability. It is best that they have a clear depth of at least 24 inches (610 mm). Check all shelving and hanging rods for adequate bracing. Check for proper type and location of closet light fixtures; lights positioned close to shelves present both a hazardous condition and an impediment to the use of shelves.

- **Trim and finishes.** Examine baseboards, sills, moldings, cornices, and other trim for missing or damaged sections or pieces. Replacement trim may no longer be readily obtainable, so determine if trim can be salvaged from more obscure locations in the building.

- **Convenience outlets and lighting.** Look for signs of inadequate or unsafe electrical service as described here and in Chapter 5. Generally speaking, each wall should have at least one convenience outlet and each room should have one switch-operated outlet or overhead light. Examine the condition of outlets and switches and feel them for overheating. Make sure they are mounted on outlet boxes and that light fixtures are securely attached to walls or ceilings.

Operate switches and look for dimmed or flickering lights that indicate electrical problems somewhere in the circuit.

The electrical system will be re-examined more thoroughly later in the inspection. Also check the light switches for sparks (arcing) when switches are turned on and off. Switches that are worn should be replaced.

- **HVAC source.** As described here and in Chapter 7, locate the heating, cooling, or ventilating source for every room. If there is a warm air supply register but no return, make sure doors are undercut one inch (25 mm) for air flow.

  With the HVAC system activated, check the heat source in each room and make sure it is functioning. The HVAC system will be more completely examined later in the inspection.

- **Skylights.** Examine the undersides of all skylights for signs of leakage and water damage. Inspect skylight components for damage, deterioration, and weathertightness. Operate openable skylights to determine their smoothness of operation, fit, and apparent weathertightness.

3.3 Bathrooms

Examine bathrooms in accordance with the procedures for other interior rooms, and additionally inspect:

- **Electrical service.** Wherever possible, switches and outlets should not be within arm's reach of the tub or shower. Consider installing ground fault interrupters (GFIs) in the outlets. See Chapter 5.

  Check the condition and operation of all switches, outlets, and light fixtures.

  If there is an exhaust fan, check its operation. It should be properly ducted to an attic vent or the building's exterior.

- **Plumbing.** Examine all exposed plumbing parts for leaking or signs of trouble or deterioration. Inspect the lavatory for secure attachment and support. Check the operation of all fixtures and decide which fixtures and trim should be replaced.

  Check the condition of all plumbing fixtures by examining for chipping, scratches, mold, stains, and other defects.

  Check the condition and operation of the lavatory, toilet, tub, and shower.

  A common problem in bathrooms is leakage around tubs and showers. If possible, inspect the ceiling below each bathroom for signs of water damage or recent patching and painting.

  Whirlpool baths should be operated for at least 20 minutes. Check how well a constant water temperature is maintained. Examine jets for evidence of mold and mildew and determine if the piping should be flushed out.

  Determine the flushing capacity of the toilet. If it is not a water-saving fixture, consider replacing it with a water-saving toilet with a 1.6 gallon (6 L) flushing capacity. Operate the toilet. Assess its bowl cleaning ability, especially if it is a water-saving fixture.

  Pressure assisted toilets use water pressure to compress air in a tank that makes the 1.5 to 1.6 gallon (5.7 to 6 L) flush very effective in cleaning the fixture bowl and preventing buildup in the soil pipe. Operate the toilet. Listen for excessive noise from vibration due to loose pieces of equipment, check for leaks, and look for rust on the tank and piping.

  Check for a faulty shower pan by covering the shower drain tightly and filling the shower base with about an inch of water. Let stand for at least an hour, if possible. Look for signs of water leakage on the ceiling below. The presence of excessive sealant around the shower base or drain may indicate attempts to remedy a shower pan leak by preventing water from reaching the pan. This is only a temporary solution and the pan should be properly repaired.

  If there is a medicine cabinet, check its condition and check its door fit and operation.

- **Tub and shower enclosures.** Check the condition, fit, and operation of tub and shower enclosures. Note whether any glazing is safety glazing, as required by the building code for new installation and replacement.

- **Ceramic tile.** Look for damaged or missing tiles, or tiles that have been scratched, pitted, or dulled by improper cleaning. Check the condition

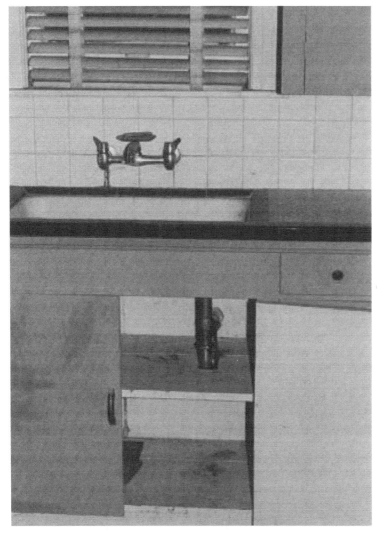

This kitchen cabinetry is plain but adequate.

3.4 Kitchens

Examine kitchens in accordance with the inspection procedures for other interior rooms, and additionally inspect:

■ **Counters and cabinetry.** Check countertops for cracks or food traps and examine kitchen cabinets carefully for signs of vermin infestation. Look for missing, broken, or damaged hardware and cabinet parts. Check doors and drawers for fit and smooth operation, and wall cabinets for secure attachment. Compare the cost of replacement to the cost of reconditioning.

■ **Electrical service.** Determine the adequacy and safety of electrical service to the kitchen, as described here and in Chapter 5. As a guide, new residential buildings are usually required to have a ground fault interrupter (GFI) of at least one 20 amp/120 volt circuit in all outlets over a countertop used for portable kitchen appliances. Separate circuits are also required for each major appliance as follows:

Refrigerator	20 amp/120 volt
Dishwasher	20 amp/120 volt
Garbage disposal	20 amp/120 volt
Range	40 to 50 amp/ 240 volt

of all grouted and caulked joints. If a portion of the tile is defective or missing, all tile may have to be replaced since finding additional tiles of matching size, color, and texture may be impossible.

■ **Ventilation.** The bathroom should be ventilated by either a window, an exhaust fan, or a recirculation fan. Poor ventilation will be indicated by mildew on the ceiling and walls.

Operate all electrical appliances simultaneously, including exhaust fans, to determine that they are connected and can run steadily without overloading their circuits.

- **Plumbing.** Visually inspect the condition of the sink for chipping, scratches, stains, and other defects. Decide whether it should be replaced. Check faucets for corrosion and proper operation. Make sure an air gap exists between the faucet and the flood rim to prevent possible back-siphoning. See Chapter 6.

   Turn the faucets on and off several times and look for drips and leaks in both the supply and drainage lines. Fill the sink and check that it drains promptly. Operate the disposal and dishwasher listening and watching for smooth operation. Look for leaks in plumbing connections. Check for the existence of an air vent for the dishwasher unless there is no disposal or unless the dishwasher pumps to the top outlet of the disposal. Check the spray hose. Decide whether either appliance should be replaced.

- **Ventilation.** See that exhaust fans and range hoods are ducted to the outside and not to a cupboard, attic, crawl space, or wall. A recirculation range hood fan is acceptable. Check the filter medium. Ducts, hoods, and filters should be free of grease buildup.

   Operate exhaust fans and vented range hoods to determine whether they are functional and whether they should be kept or replaced.

3.5 Storage Spaces

Inspect all closets and other storage spaces for cleanliness, functionality, proper lighting, and means of adequate ventilation.

The rotted supports beneath this stair were repaired by scabbing on small blocks of wood, which is insufficient. The supports should be replaced and the source of moisture investigated.

3.6 Stairs and Hallways

Inspect stairs and hallways as follows:

- **Light.** Stairs and hallways should be well lighted and have three-way light controls. Public stair and hallway lights in multifamily buildings should be operated from centralized house controls.

  Check the operation of all stair and hallway lights.

- **Smoke detectors.** Stairs and hallways are the appropriate location for smoke detectors. Detectors should be located on or near the ceiling, near the heads of stairs, and away from the corners.

  Check the operation of all smoke detectors by activating them with a smoke source or by pushing their test buttons.

- **Stair handrails and guardrails.** Handrails are normally required to be 34 to 38 inches (865 to 965 mm) above the stair nosing on at least one side of all stairs with three or more risers. Guardrails are required on open sides of stairways and should have intermediate rails that will not allow the passage of an object 4 inches (100 mm) in diameter. Shake all railings vigorously to check their stability and inspect their fastenings.

- **Stair treads and risers.** Check that all treads are level and secure. Riser heights and tread depths should be, respectively, as uniform as possible. As a guide, stairs in new residential buildings must have a maximum riser of 7-3/4 inches (197 mm) and a minimum tread of 10 inches (254 mm). Inspect the condition and fastening of all stair coverings.

- **Stair width and clearance.** Stairs should normally have a minimum headroom of 6'-8" (2030 mm) and width of 3'-0" (915 mm). For multifamily buildings, check the local housing code for minimum dimensions of public hallways and stairs.

- **Structural integrity of stairs.** Check that all stairs are structurally sound. Examine basement stairs where they meet the floor and where they are attached to the floor joists above. See Section 4.7.

3.7 Laundries and Utility Rooms

Laundry areas and utility rooms in small residential buildings are usually located in the basement or off the kitchen. Inspect them as follows:

- **Laundries.** Look for leaks or kinks in plumbing connections to the washer and examine electrical or natural gas connections to the dryer. Inspect dryer venting and make sure it exhausts to the outside and is not clogged or otherwise restricted. Gas dryer vents that pass through walls or combustible materials must be metal.

  Examine the laundry tub, if one exists, and decide whether it should be replaced. Check its plumbing and its capacity to handle discharged water from the washer.

  In multifamily laundry areas, examine floors and walls for water damage. The laundry should have a floor drain. Determine whether the laundry is of proper size and in the proper location for the planned rehabilitation.

  Operate washers and dryers and observe their functioning. Listen for noise that indicates excessive wear. Determine whether they should be replaced.

- **Furnace rooms.** Rooms containing fuel-burning equipment should not be located off a sleeping room in a single family residence, and must be in a publicly accessible area in a multifamily building. Check local code requirements for applicable fire safety and combustion air criteria.

3.8 Fireplaces and Flues

Inspect fireplaces and flues as follows:

- **Fireplaces.** Inspect the firebox for deterioration or damage. If there is a damper, check its operation. Make sure the hearth is of adequate size to protect adjacent combustible building materials, if any. A depth from the face of the fireplace of 20 inches (510 mm) and a width that extends one foot (305 mm) beyond the fireplace opening on either side is a minimum for older fireplaces. Also check local codes.

Burn some newspaper to check the draw. Discoloration around the mantel may indicate a smoky fireplace with poor draw.

- **Flues.** Check the flue lining in masonry chimneys. It should be tight along its entire length. Linings should be intact, unobstructed, and appropriate for the fuel type. It is difficult to properly examine flue linings visually, and a mirror may be helpful. An obstructed flue can usually be opened by a chimney sweep, but consult a chimney expert if the integrity of the flue is in doubt. Analyze unlined chimneys for the possible installation of metal liners. If there is an attic, use it to examine chimney construction more closely. See Section 7.2 for clearances around smoke pipes.

- **Smoke pipe connections.** Check that the smoke pipes from furnaces, water heaters, stoves, and related devices are tightly connected to the chimney and that they do not enter a fireplace flue. See additional requirements in Section 7.2.

- **Ash dump and pit.** If the fireplace has an ash dump at the bottom of the firebox, check the operation, fit, and condition of the door and check the shaft to the ash pit to be certain it is unobstructed and not overflowing with ashes. If the chimney has an ash pit, check the operation, fit, and condition of the pit access door. The fit should be tight enough to prevent dust and ash from escaping.

The structural condition of chimneys should be inspected in accordance with Section 4.6.

3.9 Attics and Roof Truss and Joist Spaces

Attics are defined here as unconditioned spaces between the roof and the ceiling or walls of the building's inhabited rooms. In small residential buildings with pitched roofs, attics are usually partially or fully accessible. In buildings with low-slope roofs, they may be inaccessible or virtually nonexistent. Inspect all accessible attic spaces as follows:

- **Roof leaks.** Look for signs of water leakage from the roof above and try to locate the source of leakage by tracing its path. This may be difficult to do beneath built-up roofs or beneath loosely laid and mechanically fastened single-ply roofs, since water may travel horizontally between layers of roofing materials. Determine the extent of any damage and the probable cost of repairs.

- **Attic ventilation.** Signs of inadequate ventilation are rusting nails (in roof sheathing, soffits, and drywall ceilings), wet or rotted roof sheathing, and excessive heat buildup in attics. Where attics exist but are inaccessible, check if access could be provided through ceilings or gable ends. Check for adequate attic ventilation by calculating the ratio of the free area of all vents to the floor area. Free area of vents is their clear, open area. If a vent has an insect screen, its free area is reduced by half. The vent free area to floor area ratio should be 1 to 150. If the calculated ratio is less, consider adding ventilation, especially in hot and humid climates. When an attic also contains an occupied space, check that the ventilation from the unconditioned, unoccupied areas at the eaves is continuous to the gable or ridge vents. Also check that the free area of eave vents is approximately equal to the free area of ridge or gable vents. If ventilation appears to be inadequate and additional vents cannot be added economically, consider adding mechanical ventilation.

- **Roof truss and joist space ventilation.** Most buildings with low-slope roofs and some buildings with pitched roofs do not have attics. Instead, these buildings have ceilings at the bottom of joists, rafters, or trusses. The truss space and the space between each joist or rafter and above the ceiling needs ventilation. Look for vents below the eaves and check to see that the ratio of free vent area to roof area is 1 to 150. If the calculated ratio is less, consider adding ventilation. Open one or more of the vents if possible. Probe the ventilating cavity to determine the amount of insulation and free air space and try to assess the general condition of the surrounding building components. It is difficult to inspect for ventilation in these

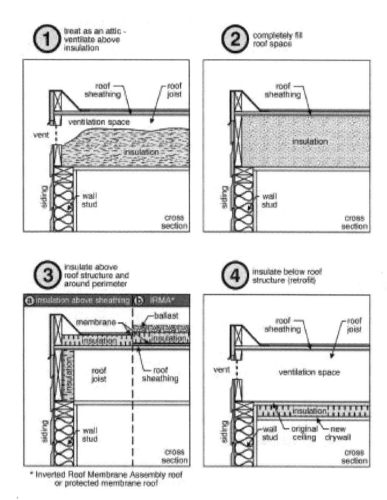

Methods of insulating and ventilating low-slope roof structures

buildings without removing a part of the ceiling to measure the free depth and width of ventilation space and to determine whether the truss, joist, or rafter spaces contain insulation. If there is no evidence of water damage from condensation, an intrusive investigation is usually not warranted. At ridge, cornice, eave, or soffit vents, check for the presence of insect screens. If insect screens are present, check their condition.

- **Vents.** Check the condition of ridge, gable, cornice, eave, and soffit vents. Look for rusted or broken screens and rusted frames. Make sure openings are clear of dirt and debris. At larger ventilation openings on a building's exterior and where louvered grilles are used, such as at gables, check for the presence of one-half-inch-square (13 x 13 mm) 14 or 16 gauge aluminum mesh bird screen. If there is none or it is in poor condition, consider having new bird screen installed.

- **Thermal insulation.** Examine the amount and type of existing insulating material. Check to see that insulation faced with a vapor barrier has been installed face-side down with the vapor barrier closest to the conditioned space, and that vapor barriers are properly located between the ceiling and the first layer of insulation. Determine the proper amount of insulation to the attic and whether additional insulation is needed. If attic insulation is placed against the roof sheathing, check for a ventilated air space between the insulation and the sheathing. If there is no air space, check for the presence of moisture and deterioration of sheathing and rafters. Ensure that insulation is held away from recessed lighting fixtures, and inspect spaces around vents, stacks, ducts, and wiring for thermal bypasses. Inspect attic doors or access hatches, heating or cooling ducts that pass through the attic, and whole-house attic fans for thermal bypasses. Check the local jurisdiction for thermal resistance (R) requirements.

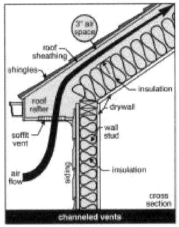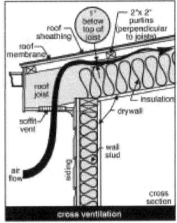

for cathedral ceilings and flat roofs the recommended vent area is 1 square foot for every 150 square feet of roof area

open web trusses also permit cross ventilation

Methods of ventilating pitched roof structures over habitable spaces

A split ceiling joist or trust chord can be easily repaired by propping it back in place and attaching new structural pieces to each side.

- **Exhaust ducts and plumbing stacks.** Check that all plumbing stacks continue through the roof and do not terminate in the attic and that they are not broken or damaged. Also check that exhaust ducts are not broken or damaged and do not terminate in the attic but either continue through the roof, gable, or wall or vent directly to a ridge vent.
- **Structural conditions.** Inspect the roof structure in accordance with Section 4.7.

3.10 Whole Building Thermal Efficiency Tests

Several whole building tests can be performed to help evaluate the thermal efficiency of the building envelope.

Test: A building pressurization test can be used to determine air infiltration and exfiltration. The test is particularly useful for "tightening up" an older building. See ASTM E779, *Standard Test Method for Determining Air Leakage Rate by Fan Pressurization*. A tracer gas test may also be used; see ASTM E741, *Standard Test Method for Determining Air Change in a Single Zone by Means of a Tracer Gas Dilution*. Such tests are usually performed by an energy specialist or an HVAC technician.

Test: A hand-held infrared scanner can be used to detect building "hot spots" due to interior air leakage or excessive heat loss through uninsulated building components. This test should be performed in cold weather when the building is heated; the greater the differential between inside and outside temperatures, the more accurate the results. Infrared scanners are commercially available; their use varies by manufacturer. Thermography can be used for the same purpose, but it requires much more expensive equipment and a trained operator. Thermographic tests should be performed by an energy specialist, mechanical engineer, or others with the proper training and equipment.

3.11
Sound Transmission Control Between Dwelling Units

Check the floors and walls between dwelling units for adequacy of sound transmission control using the current building code for guidance. Floors that separate dwelling units, and floors that separate a dwelling unit from a public or service area should have an Impact Insulation Class (IIC) of not less than 45. IIC is determined in accordance with ASTM E492, *Standard Method of Laboratory Measurement of Impact Sound Transmission Through Floor-Ceiling Assemblies Using the Tapping Machine*. Walls and floors that separate dwelling units in two-family residences, and walls that separate townhouses, should have a Sound Transmission Class (STC) of not less than 45. STC is determined in accordance with ASTM E90, *Standard Method for Laboratory Measurement of Airborne Sound Transmission Loss of Building Partitions and Elements*.

Technical data that identifies STC and IIC attenuation for different types of construction is provided by product manufacturers and trade associations, such as the Gypsum Association or the National Concrete Masonry Association. See the Gypsum Association's publication GA-530, *Design Data—Gypsum Products*, and the National Concrete Masonry Association's *Tek Note 13-1, Sound Transmission Class Ratings for Concrete Masonry Walls*.

3.12
Asbestos

Asbestos is a naturally occurring fibrous mineral used in many construction products. It is considered to be a carcinogen. Asbestos has been used in sealant, putty, and spackling compounds; in vinyl floor tiles, backing for vinyl sheet flooring, and flooring adhesives; in ceiling tiles; in textured paint; in exterior wall and ceiling insulation; in roofing shingles; in cement board for many uses including siding; in door gaskets for furnaces and wood-burning stoves; in concrete piping; in paper, mill board, and cement board sheets used to protect walls and floors around wood-burning stoves; in fabric connectors between pieces of metal ductwork; in hot water and steam piping insulation, blanket covering, and tape; and as insulation on boilers, oil-fired furnaces, and coal-fired furnaces. Use of asbestos has been phased out since 1978, but many older houses contain asbestos-bearing products.

Products containing asbestos are not always a health hazard. The potential health risk occurs when these products become worn or deteriorate in a way that releases asbestos fibers into the air. Of particular concern are those asbestos-containing products that are soft, that were sprayed or troweled on, or that have become crumbly. The Environmental Protection Agency believes that so long as the asbestos-bearing product is intact, is not likely to be disturbed, and is in an area where repairs or rehabilitation will not occur, it is best to leave the product in place. If it is deteriorated, it may be enclosed, coated or sealed up (encapsulated) in place, depending upon the degree of deterioration. Otherwise, it should be removed.

A certified environmental professional should perform the inspection and make the decision whether to enclose, coat, encapsulate, or remove deteriorated asbestos-containing products. Testing by a qualified laboratory as directed by the environmental professional may be needed in order to make an informed decision. Encapsulation, removal, and disposal of asbestos products must be done by a qualified asbestos-abatement contractor.

For more information consult the *Guidance Manual: Asbestos Operations and Maintenance Work Practices*, available from the National Institute of Building Sciences.

3.13 Lead

Lead has been determined to be a significant health hazard if ingested, especially by children. Lead damages the brain and nervous system, adversely affects behavior and learning, slows growth, and causes problems related to hearing, pregnancy, high blood pressure, nervous system, memory, and concentration.

- **Lead-based paint.** Most homes built before 1940 used paint that was heavily leaded. Between 1940 and 1960, no more than half the homes built are thought to have used heavily leaded paint. In the period from 1960 to 1980, many fewer homes used lead-based paint. In 1978, the U.S. Consumer Product Safety Commission (CPSC) set the legal limit of lead in most types of paint to a trace amount. As a result, homes built after 1978 should be nearly free of lead-based paint. In 1996, Congress passed the final phase of the Residential Lead-Based Paint Hazard Reduction Act, Title X, which mandates that real estate agents, sellers, and landlords disclose the known presence of lead-based paint in homes built prior to 1978.

  Lead-based paint that is in good condition and out of the reach of children is usually not a hazard. Peeling, chipping, chalking, or cracking lead-based paint is a hazard and needs immediate attention. Lead-based paint may be a hazard when found on surfaces that children can chew or that get a lot of wear and tear, such as windows and window sills, doors and door frames, stairs, railing, banisters, porches, and fences. Lead from paint chips that are visible and lead dust that is not always visible can both be serious hazards. Lead dust can form when lead-based paint is dry scraped, dry sanded, or heated. Dust also forms when painted surfaces bump or rub together, such as when windows open and close. Lead chips and dust can get on surfaces and objects that people touch. Settled lead dust can re-enter the air when people vacuum, sweep, or walk through it.

  If the building is thought to contain lead-based paint, consider having a qualified professional check it for lead hazards. This is done by means of a paint inspection that will identify the lead content of every painted surface in the building and a risk assessment that will determine whether there are any sources of serious lead exposure (such as peeling paint and lead dust). The risk assessment will also identify actions to take to address these hazards. The federal government is writing standards for inspectors and risk assessors. Some states may already have standards in place. Call local authorities for help with locating qualified local professionals. While home test kits for lead are available, the federal government is still testing their reliability. These tests should not be the only method used before doing rehabilitation or to ensure safety.

  For more information on lead-based paint consult the HUD Office of Lead Hazard Control Web site at http://www.hud.gov:80/lea.

- **Lead in drinking water** is a direct result of lead that is part of the plumbing system itself. Lead solder was used in pipe fittings in houses constructed prior to 1988. Lead has been used in plumbing fixtures such as faucets, and in some older homes the service water pipe from the main in the street to the house is made of lead.

  The transfer of lead into water is determined primarily by exposure (the length of time that water is in contact with lead). Two other factors that affect the transfer are water temperature (hot water dissolves lead quicker than cold water) and water acidity ("soft" water is slightly corrosive and reacts with lead).

  The current federal standard for lead in water is a limit of 15 parts per billion. The only way to find out whether there is lead in the house's water is to have the water tested by an approved laboratory. Two samples should be tested: water that has been sitting in the pipes for at least four hours, and water that has been allowed to flow not less than one minute before the sample is taken. Tests are inexpensive ($15 to $25).

If there is evidence of lead in the system, consider having water tested for lead. If the house has a water filter, check to see if it is certified to remove lead.

For more information on lead in drinking water call the Environmental Protection Agency's Safe Drinking Water Hotline: 1-800-462-4791 or visit the Web site of the EPA Office of Water at http://www.epa.gov/safewater.

For more information on lead hazards in general, call the National Lead Information Center clearinghouse: 1-800-424-LEAD. For the hearing impaired, call TTY 1-800-526-5456.

3.14 Radon

Radon is a colorless, odorless, and tasteless gas that is present in varying amounts in the ground and in water. Radon is produced by the natural radioactive decay of uranium deposits in the earth. Prolonged exposure to radon in high concentrations can cause cancer. The EPA has set guidelines for radon levels in residential buildings.

- **Airborne radon.** The EPA recommends that mitigation measures be undertaken in residential buildings when radon concentrations are 4 picocuries per liter (4 pCi/L) of air and above.

  The radon concentration in a house varies with time and is affected by the uranium-radium content in the soil, the geological formation beneath the house, the construction of the house, rain, snow, barometric pressure, wind, and pressure variations caused by the periodic operation of exhaust fans, heating systems, fireplaces, attic fans, and range fans. Radon concentrations are variable and may be high in one house and low in an adjacent house. To determine if a house has a radon problem, it must be tested.

  Test: A long-term test is the most accurate method of determining the average annual radon concentration. However, because time is usually limited, there is a three- to seven-day test that uses a charcoal canister. It is available from most home do-it-yourself stores or through radon testing service companies.

- **Waterborne radon.** A house's domestic water supply from its well can contain radon. There are locations with well water containing 40,000 or more pCi/L. The health problems from drinking water with radon are insignificant compared to breathing airborne radon, but radon can be released into the air when water is run into a plumbing fixture or during a shower. It takes a high concentration of radon in water to produce a significant concentration in the large volume of air in a house. While there is no maximum established at this time for radon in water, consider removing radon at the water service entrance when the level exceeds 10,000 pCi/L.

  Test: Private well water testing is normally not a part of radon testing. Therefore, if the house has a private well, consult the local health department to determine whether water testing in the house's area is recommended. How this is to be done should be determined by a listed radon mitigation contractor.

If a building is found to have a radon problem, consult a certified radon mitigation contractor who has met the requirements for listing under the EPA's Radon Contractor Proficiency Program about mitigation procedures.

For more information on radon mitigation see Ventilation for Soil Gases, chapter 7 in *The Rehab Guide*, Volume 1, *Foundations*, available from HUD or full text online at http://www.pathnet.org.

3.15 Tornado Safe Room

If a building is located in a tornado-risk area, and if it has a tornado shelter or safe room, it should be investigated by a structural engineer for structural adequacy.

4
Structural System

Small residential buildings do not have for the most part inherent structural problems. Even many 19th-century buildings, that often show signs of settlement, may have only minor structural faults that can be readily remedied. Major structural problems, when they do occur, are usually quite obvious. It is the less obvious problems that require careful inspection and informed diagnosis. Such problems are often detected through a pattern of symptoms rather than any one symptom. This chapter describes signs of structural distress, deterioration, and damage by material type: masonry (the most difficult to assess), wood, iron and steel, and concrete. If applicable, also refer to Appendix A, The Effects of Fire on Structural Systems, and Appendix B, Wood Inhabiting Organisms.

4.1
Seismic Resistance

If the building is in seismic zones 2B, 3, and 4 (California, Idaho, Nevada, Oregon, Washington, and portions of Alaska, Arizona, Arkansas, Hawaii, Missouri, Montana, New Mexico, Utah, and Wyoming), have a structural engineer check the following conditions for structural vulnerability. Note that wood frame buildings with brick or stone veneer are still considered wood frame.

Figure 4.1
Assessing Structural Capacity

Unless there is obvious overloading, significant deterioration of important structural components, or additional loading is anticipated, there is usually little need to verify the building's structural design or to recompute its structural capacity. A thorough visual inspection of its structural components is all that is normally necessary. On some occasions, however, ceilings or other building elements may have to be opened for a selective inspection of critical structural members. A sufficient number of members must be examined to afford a reasonable assurance that they are representative of the total structure.

Test: A laboratory test may be helpful in determining the strength of a masonry wall section or some other structural component. A representative sample of the material or component in question must be removed from the structure for the test, which should be performed under the guidance of a qualified structural engineer and conducted by a certified testing laboratory. Refer to the ASTM tests listed in Sections 4.3, 4.8, and 4.9.

If doubt remains about the building's structural capacity after visual inspection, laboratory tests may be the next step. Also, if the building's structural loading is to be dramatically increased during its rehabilitation by such things as a water bed, stone kitchen countertop, tile flooring, or heavy stove and oven, a quantitative analysis should be made of all the structural members involved. Simple calculations may be made in accordance with *Ambrose's Simplified Engineering for Architects and Builders* or, when wood-supported spans are involved, the use of the span tables in CABO *One- and Two-Family Dwelling Code*, the *International Residential Code for One- and Two-Family Dwellings*, or the local building code may be sufficient. More complex calculations should be performed by a qualified structural engineer in accordance with ASCE 11, *Guideline for Structural Condition Assessment of Existing Buildings*.

- Wood frame buildings that are **not physically anchored** to their foundations. Such buildings may be vulnerable to shifting or sliding.

- Wood frame buildings and wood-framed portions (porches, for example) or other buildings when they are **supported above ground** on either short wood studs (cripple walls) or on piers of stone, masonry, or concrete. Such buildings may be vulnerable to tilting or falling over.

- **Unreinforced and inadequately reinforced masonry buildings.** Such buildings may be vulnerable to total or partial collapse due to inadequate reinforcement or to inadequate anchorage of roofs and walls to the floors. Use as a reference Seismic Strengthening Provisions for Unreinforced Masonry Bearing Wall Buildings, Appendix Chapter 1 of the *Uniform Code for Building Conservation*.
- **Buildings of any type that have irregular shapes.** Such buildings may be vulnerable to partial collapse.
- **Wood frame and masonry buildings with more than one story above grade where the story at grade is a large unobstructed open space,** such as a garage. Such buildings may be vulnerable to collapse of the story at grade.
- **Wood frame and masonry buildings with more than one story above grade that are constructed on sloping hillsides,** and buildings of any type of construction and height that are constructed on steep slopes of 20 °F (-7 °C) or more. Such buildings may be vulnerable to sliding.

If the building is in seismic zone 2A (Connecticut, Massachusetts, Rhode Island, South Carolina, and portions of Georgia, Illinois, Indiana, Kansas, Kentucky, Maine, New Hampshire, New Jersey, New York, North Carolina, Oklahoma, Pennsylvania, Tennessee, Vermont, and Virginia) and has more than two stories above grade, consider having a structural engineer check for the last two conditions (large unobstructed open space at grade and sloped sites).

Buildings not of wood frame or masonry construction, such as stone, adobe, log, and post and beam structures, as well as buildings with more than one type of construction in any seismic zone, should be investigated by a structural engineer to determine their seismic vulnerability.

Masonry bearing wall buildings in seismic zones 2B, 3, and 4 should be investigated by a structural engineer for the presence of reinforcing steel. Use as a reference Seismic Strengthening Provisions for Unreinforced Masonry Bearing Wall Buildings, Appendix Chapter 1 of the *Uniform Code for Building Conservation*.

Have a structural engineer check the anchorage of wood framed structures to their foundations and investigate all such structures supported on cripple walls or piers in seismic zones 2, 3, and 4.

In all seismic zones, a structural engineer should investigate buildings with more than one story above grade where the story at grade is a large unobstructed open space or the building is on a sloping hillside and in seismic zones 2B, 3, and 4, buildings with an irregular shape.

4.2 Wind Resistance

The coasts of the Gulf of Mexico, the south- and mid-Atlantic coast, the coastal areas of Puerto Rico, the U.S. Virgin Islands and Hawaii as well as the U.S. territories of American Samoa and Guam are vulnerable to hurricanes in the late summer and early fall. Hurricanes are large, slow moving, damaging storms characterized by gusting winds from different directions, rain, flooding, high waves, and storm surges. Winter storms along the mid- and north-Atlantic coast can be more damaging than hurricanes because of their greater frequency, longer duration, and high erosion impacts on the coastline. Even in states not normally considered susceptible to extreme wind storms, there are areas that experience dangerously high winds. These areas are typically near mountain ranges and include the Pacific Northwest coast. Other extreme wind areas include the plains states, which are especially subject to tornadoes.

In addition to the direct effects of high winds and winter on buildings, hurricanes and other severe storms generate airborne debris that can damage buildings. Debris, such as small stones, tree branches, roof shingles and tiles, building parts and other objects, is picked up by the wind and moved with enough force to damage and even penetrate windows, doors, walls, and roofs. When a building's exterior envelope is breached by debris, the building can become pressurized, subjecting its walls and roof to much higher damaging wind pressures. In general, the stronger the wind, the larger and heavier the debris it can carry and the greater the risk of severe damage.

If the building is in a hurricane or high-wind region, have a structural engineer check its structural system for continuity of load path, including resistance to uplift forces. If there is an accessible attic, check for improper attachment of the roof sheathing to the roof framing members by looking for unengaged or partially engaged nails. Check for the presence of hurricane hold-down clips for joists, rafters, and trusses at the exterior wall. Examine the gable end walls and the roof trusses for lateral bracing. Check to see whether the exterior wall and other load-bearing walls are securely attached to the foundation.

A common masonry wall crack probably caused by thermal or moisture expansion. If possible, monitor such cracks over a period of time to see if they're active. Active cracks should be sealed with a flexible sealant; inactive cracks may be pointed.

4.3 Masonry, General

All exposed masonry should be inspected for cracking, spalling, bowing (bulges vertically), sweeping (bulges horizontally), leaning, and mortar deterioration. Before beginning a detailed masonry inspection, determine which walls are load-bearing and which are not. Usually this can be done by examining the beams and joists in the building's basement or crawl space or attic. Note also whether the walls are solid masonry or masonry cavity, or whether they are non-structural brick or stone veneer. The overall quality of the building's construction, and often that of its neighborhood, will be a good indicator of the condition of its masonry.

There may be a substantial difference in the masonry walls in buildings built during the last 40 to 50 years compared to those constructed earlier. Walls became thinner as designers began to exploit more effectively the compressive strength of masonry by using higher strength masonry materials and mortars. This change came at the expense of flexibility as such walls are often more brittle than their massive ancestors and, therefore, more subject to stress-induced damage.

Tests: Two methods of testing are sometimes useful for assessing masonry. Probe holes can be drilled through the joints or masonry units with a masonry bit and probed with a stiff wire (or, if available, a fiber optic device) to determine a wall's thickness and the adequacy of its mortar. The probe holes are patched after the investigation has been completed. A hammer test can be used to determine the structural soundness of masonry units and their bond to the mortar. The masonry is tapped lightly with a hammer and the resonance of the sound produced is evaluated. Two tests may also be useful: ASTM E518, *Standard Test Methods for Flexural Bond Strength of Masonry*, and ASTM E519, *Standard Test Method for Diagonal Tension (Shear) in Masonry Assemblages*. See a qualified masonry consultant for the proper use of these tests on existing masonry.

The brick shown here is highly spalled from the effects of excessive moisture penetration and subsequent freezing. The damage cannot be repaired, although individual bricks can be replaced and the mortar pointed. The new mortar should be of the same composition as the old.

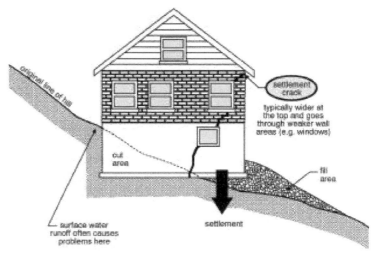

Building settlement due to cut and fill excavation

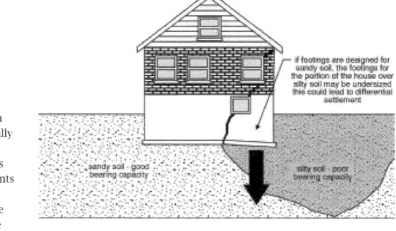

Differential settlement caused by variable soil types

- **Masonry cracking.** Although masonry can deform elastically over long periods of time to accommodate small amounts of movement, large movements normally cause cracking. Cracks may appear along the mortar joints or through the masonry units. Cracking can result from a variety of problems: differential settlement of foundations, drying shrinkage (particularly in concrete block), expansion and contraction due to ambient thermal and moisture variations, improper support over openings, the effects of freeze-thaw cycles, the corrosion of iron and steel wall reinforcement, differential movement between building materials, expansion of salts, and the bulging or leaning of walls. These problems are more fully discussed in Sections 4.4 and 4.5.

  Cracks should always be evaluated to determine their cause and whether corrective action is required. Look for

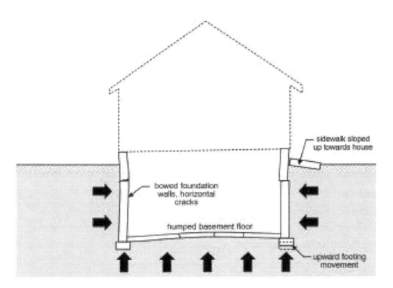

Evidence of frost heaving

An extreme case of structural failure in a masonry wall due to foundation settlement. The wall and foundation must be completely rebuilt.

signs of movement. A clean crack indicates recent movement; a dirty or previously filled crack may be inactive (a pocket lens may be useful for such an examination). Correlate the width of larger cracks to the age of the building. A one-half-inch crack in a new building may be a sign of rapid settlement, but in a building 50 years old, it may indicate a very slow movement of only 1/100 of an inch (0.25 mm) per year. In each case the cause and treatment may differ.

Test: Crack movement can be measured with a commercially available joint movement indicator. This device is temporarily fastened over the crack and a scribe records movement over a period of time. Cyclical movements may take six months or more to measure, but diurnal movements can be recorded over a few days. Hand measurements can also be made of crack movements, but these will be less precise and require repeated field visits.

Cracks associated with thermal expansion and contraction may open and close with the season. These are cyclical cracks, which may gradually expand as accumulating mortar debris jams them farther apart after each cycle. Such cracks should be cleaned and protected by flexible sealants; remortaring cyclical cracks will hold them open and cause more cracking.

When there are masonry problems, it is advisable to procure the services of a structural engineer. If problems appear to be due to differential settlement, a soils engineer also may be required.

- **Mortar deterioration.** The two important qualities of mortar are its ability to bond to masonry and its internal strength. A sign of poorly made mortar may be random cracking at the bond joint. Until about the end of the 19th century, the standard mortar for masonry was a mixture of sand and pure lime or lime-possolan-sand. These low strength mortars gave masonry the ability to absorb considerable strain. Accordingly, the tendency to crack was reduced and when cracks did appear in the mortar joints, they were to a great extent capable of chemical reconstitution or "self healing." Thus, the age of the building may be a good clue in evaluating its mortar problems. Older mortar (or mortar of any age that uses hydrated lime) will be softer and may require pointing, but otherwise may be responsible for a sound wall.

Most often, mortar deterioration is found in areas of excessive moisture, such as near leaking downspouts, below windows, and at tops of walls. In such cases the remedy is to redirect the water flow and point the joints. Pointing should be performed with mortar of a composition similar to or compatible with the original mortar. The use of high strength mortar to point mortar of a lower strength can do serious damage to the masonry since the pointing can't "flex" with or act in a similar way to the rest of the joint. It is useful to remember that mortar acts as a drainage system to equalize hydrostatic pressure within the masonry. Nothing should be done to reduce its porosity and thereby block water flow to the exterior surface.

Test: To determine the composition (percentage of lime and other materials) of existing mortar, remove a sample and have it chemically analyzed by a testing laboratory. This should be done under the supervision of a qualified structural engineer.

- **Deterioration of brick masonry units.** The spalling, dusting, or flaking of brick masonry units may be due to either mechanical or chemical damage. Mechanical damage is caused by moisture entering the brick and freezing, resulting in spalling of the brick's outer layers. Spalling may continue or may stop of its own accord after the outer layers that trapped the interior moisture have broken off. Chemical damage is due to the leaching of chemicals from the ground into the brick, resulting in internal deterioration. External signs of such deterioration are a dusting or flaking of the brick.

Very little can be done to correct existing mechanical and chemical damage except to replace the brick. Mechanical deterioration can be slowed or stopped by directing water away from the masonry surface and by pointing mortar joints to slow water entry into the wall. Surface sealants (damp proofing coatings) are rarely effective and may hasten deterioration by trapping moisture or soluble salts that inevitably penetrate the wall and in turn cause further spalling. Chemical deterioration can be slowed or stopped by adding a damp proof course (or injecting a damp proofing material) into the brick wall just above the ground line. Consult a masonry specialist for this type of repair.

4.4 Masonry Foundations and Piers

Inspect stone, brick, concrete, or concrete block foundations for signs of the following problems (this may require some digging around the foundation):

- **Problems associated with differential settlement.** Uneven (differential) settlement can be a major structural problem in small residential buildings, although serious settlement problems are relatively uncommon. Many signs of masonry distress are incorrectly diagnosed as settlement-related when in fact they are due to moisture and thermal movements.

Indications of differential settlement are vertical distortion or cracking of masonry walls, warped interior and exterior openings, sloped floors, and sticking doors and windows. Settlement most often occurs early in the life of a building or when there is a dramatic change in underground conditions. Often such settlement is associated with improper foundation design, particularly inadequate footers and foundation walls. Other causes of settlement are:

❏ *soil consolidation* under the footings

❏ *soil shrinkage* due to the loss of moisture to nearby trees or large plants

❏ *soil swelling* due to inadequate or blocked surface or house drainage

❏ *soil heaving* due to frost or excessive root growth

❏ *gradual downward drift of clay* soils on slopes

❏ *changes in water table* level

❏ *soil erosion* around footers from poor surface drainage, faulty drains, leaking water mains or other underground water movements (occasionally, underground water may scour away earth along only one side of a footer, causing its rotation and the subsequent buckling or displacement of the foundation wall above)

❏ *soil compaction* or movement due to vibration from heavy equipment, vehicular traffic, or blasting, or from ground tremors (earthquakes).

This pier has been overstressed by movement of the porch and column. The entire assembly should be rebuilt.

Gradual differential settlement over a long period of time may produce no masonry cracking at all, particularly in walls with older and softer bricks and high lime mortars; the wall will elastically deform instead. More rapid settlements, however, produce cracks that taper, being largest at one end and diminishing to a hairline at the other, depending on the direction and location of settlement below the wall. Cracking is most likely to occur at corners and adjacent to openings, and usually follows a rough diagonal along mortar joints (although individual masonry units may be split). Settlement cracks (as opposed to the similar-appearing shrinkage cracks that are especially prevalent in concrete block) may extend through contiguous building elements such as floor slabs, masonry walls above the foundation, and interior plaster work. Tapering cracks, or cracks that are nearly vertical and whose edges do not line up, may occur at the joints of projecting bay windows, porches, and additions. These cracks indicate differential settlement due to inadequate foundations or piers under the projecting element.

Often settlement slows a short time after construction and a point of equilibrium is reached in which movement no longer occurs. Minor settlement cracking is structurally harmful only if long-term moisture leakage through the cracks adversely affects building elements. Large differential settlements, particularly between foundation walls and interior columns or piers, are more serious because they will cause movements in contiguous structural elements such as beams, joists, floors, and roofs that must be evaluated for loss of bearing and, occasionally, fracture.

Should strengthening of the foundation be required, it can be accomplished by the addition of new structural elements, such as pilasters, or by pressure-injecting concrete epoxy grout into the foundation wall. If movement continues and cracking is extensive, it is possible that the problem can only be rectified by underpinning. Older buildings with severe settlement problems may be very costly to repair. Seek the advice of a structural or soils engineer in such cases.

■ **Problems associated with masonry piers.** Masonry piers are often used to support internal loads on small residential buildings or to support

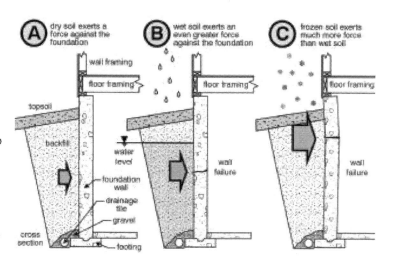

The effect of soil pressure on foundation walls

projecting building elements such as bay windows, porches, and additions. In some cases they support the entire structure. Piers often settle differentially and over a long period of time (particularly when they are exposed to the weather) they tend to deteriorate. Common problems are:

❏ *Settlement or rotation of the pier footing,* which causes a lowering or tilting of the pier and subsequent loss of bearing capacity. Wood frame structures adjust to this condition by flexing and redistributing their loads or by sagging (see Section 4.7). Masonry walls located over settled piers will crack.

❏ *Frost heaving of the footing or pier,* a condition caused by the lack of an adequate footing or one of insufficient depth. This will result in raising or tilting the pier, and in structural movement above it similar to that caused by settlement or rotation of the footing. Such a condition is most common under porches.

❏ *Physical deterioration of the pier* due to exposure, poor construction, or overstressing. Above-ground piers exposed to the weather are subject to freeze-thaw cycles and subsequent physical damage. Piers for many older residential structures are often of poorly constructed masonry that deteriorates over the years. A sign of overstressing of piers is vertical cracking or bulging.

❏ *Loss of bearing of beams, joists, or floors* due to the above conditions or due to movements of the structure itself.

Piers should be examined for plumbness, signs of

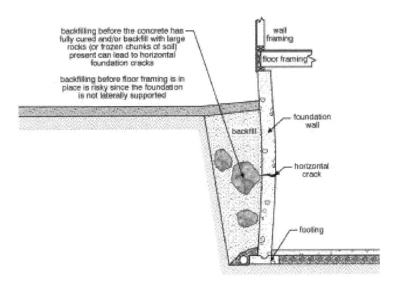

How backfilling can affect foundation walls

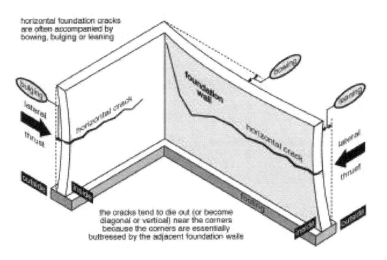

How horizontal cracks relate to foundation wall movement

settlement, condition, and their adequacy in accepting bearing loads. Check their width to height ratio, which should not exceed 1:10. Those that are deficient should be repaired or replaced. When appearance is not a factor (as is often the case), piers can be supplemented by the addition of adjacent supports.

■ **Cracking associated with drying shrinkage in concrete block foundation walls.** The shrinkage of concrete block walls as they dry in place often results in patterns of cracking similar to that caused by differential settlement: tapering cracks that widen as they move diagonally upward. These cracks usually form during the building's first year, and in existing buildings will appear as "old" cracks and exhibit no further movement. Although such cracks are often mistaken for settlement cracks, shrinkage cracks usually occur in the middle one-third of the wall and the footer beneath them remains intact. Shrinkage cracking is rarely serious, and in an older building may have been repaired previously. If the wall is unsound, its structural integrity sometimes can be restored by pressure-injecting concrete epoxy grout into the cracks or by adding pilasters.

■ **Sweeping or horizontal cracking of the foundation walls.** The sweeping or horizontal cracking of brick or concrete block foundation walls may be caused by improper backfilling, vibration from the movement of heavy equipment or vehicles close to the wall, or by the swelling or freezing and heaving of water saturated soils adjacent to the wall. Like drying shrinkage, sweeping or horizontal cracking may have occurred during the original construction and been compensated for at that time. Such distress, however, is potentially serious as it indicates that the vertical supporting member (the foundation wall) that is carrying a portion of the structure above is "bent" or "broken." It may be possible to push the wall back into place by careful jacking, and then reinforcing it with the addition of interior buttresses or by pressure-injecting concrete epoxy grout into the wall. If outside ground conditions allow, the wall can be relieved of some lateral pressure by lowering the ground level around the building. When expansive soils are suspected as the cause of the cracking, examine the exterior for sources of water such as broken leaders or poor surface drainage. Suspect frost heaving if the damage is above local frost depth or if it occurred during an especially cold period.

4.5 Above-Ground Masonry Walls

Inspect above-ground stone, brick, or concrete block walls for signs of the following problems:

■ **Brick wall cracking associated with thermal and moisture movement.** Above-ground brick walls expand in warm weather (particularly if facing south or west) and contract in cool weather. This builds up stresses in the walls that may cause a variety of cracking patterns, depending on the configuration of the wall and the number and location of openings. Such cracks are normally cyclical and will open and close with the season. They will grow wider in cold weather and narrower in hot weather. Look for cracking at the corners of long walls, walls with abrupt changes in cross section (such as at a row of windows), walls with abrupt turns or jogs, and in transitions from one- to two-story walls. These are the weak points that have the least capacity for stress. Common moisture and thermal movement cracking includes:

❑ *Horizontal or diagonal cracks near the ground at piers in long walls* due to horizontal shearing stresses between the upper wall and the wall where it enters the ground. The upper wall can thermally expand but its movement at ground level is moderated by earth temperatures. Such cracks extend across the piers from one opening to another along the line of least resistance. This condition is normally found only in walls of substantial length.

❑ *Vertical cracks near the end walls* due to thermal movement. A contracting wall does not have the tensile strength to pull its end walls with it as it moves inward, causing it or the end walls to crack vertically where they meet.

❑ *Vertical cracks in short offsets and setbacks* caused by the thermal expansion of the longer walls that are adjacent to them. The shorter walls are "bent" by this thermal movement and crack vertically.

❑ *Vertical cracks near the top and ends of the facade* due to the thermal movement of the wall. This may indicate poorly bonded masonry. Cracks will tend to follow openings upward.

❑ *Cracks around stone sills or lintels* caused by the expansion of the masonry against both ends of a tight-fitting stone piece that cannot be compressed.

Cracks associated with thermal and moisture movement often present only cosmetic problems. After their cause has been determined, they should be repaired with a flexible sealant, since filling such cyclic cracks with mortar will simply cause the masonry to crack in another location. Cracks should be examined by a structural engineer and may require the installation of expansion joints.

■ **Brick wall cracking associated with freeze-thaw cycles and corrosion.** Brick walls often exhibit distress due to the expansion of freezing water or the rusting of embedded metals. Such distress includes:

❑ *Cracking around sills, cornices, eaves, chimneys, parapets, and other elements subject to water penetration,* which is usually due to the migration of water into the masonry. The water expands upon freezing, breaking the bond between the mortar and the masonry and eventually displacing the masonry itself. The path of the water through the wall is indicated by the pattern of deterioration.

❑ *Cracking around iron or steel lintels,* which is caused by the expansive force of

Despite the loss of masonry, this arch is intact and can be repaired with matching bricks.

corrosion that builds up on the surface of the metal. This exerts great pressure on the surrounding masonry and displaces it, since corroded iron can expand to many times its original thickness. Structural iron and steel concealed within the masonry, if exposed to moisture, can also corrode, and cause cracking and displacement of its masonry cover. Rust stains usually indicate that corrosion is the cause of the problem. Check to make sure the joint between the masonry and the steel lintel that supports the masonry over an opening is clear and

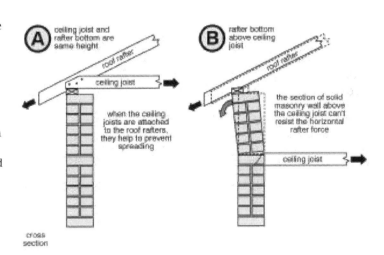

Walls that extend above ceiling joists

open. If the joint has been sealed, the sealant or mortar should be removed.

These conditions usually can be corrected by repairing or replacing corroded metal components and by repairing and pointing the masonry. Where cracking is severe, portions of the wall may have to be reconstructed. Cracks should be examined by a structural engineer.

■ **Wall cracking or displacement associated with the structural failure of building elements.** Structure-related problems, aside from those caused by differential settlement or earthquakes, are usually found over openings and (less commonly) under roof eaves or in areas of structural overloading. Such problems include:

❑ *Cracking or displacement of masonry over openings,* resulting from the deflection or failure of the lintels or arches that span the openings. In older masonry walls with wood lintels, cracking will occur as the wood sags or decays. Iron and steel lintels also cause cracking as they deflect over time. Concrete and stone lintels occasionally bow and sometimes crack.

Masonry arches of brick or stone may crack or fail when there is wall movement or when their mortar joints deteriorate.

When such lintel deflections or arch failures occur, the masonry above may be supporting itself and will exhibit step cracks beginning at the edges of the opening and joining in an inverted "V" above the opening's midpoint. Correcting such problems usually means replacing failed components and rebuilding the area above the opening.

Occasionally masonry arches fail because the walls that surround them cannot provide an adequate counterthrust to the arch action. This sometimes happens on windows that are too close to the corners of a wall or bay. In such cases, the masonry arch pushes the unbraced wall outward, causing it to crack above the opening near or just above the spring of the arch. When this occurs, the end walls must be strengthened.

❑ *Cracking or outward displacement under the eaves of a pitched roof* due to failure in the horizontal roof ties that results in the roof spreading outward. The lateral thrust of the roof on the masonry wall may cause it to crack horizontally just below the eaves or to move outward with the roof. The roof will probably be leaking as well. When this occurs, examine the roof structure carefully to ascertain whether there is a tying failure. If so, additional horizontal ties or tension members will have to be added and, if possible, the roof pulled back into place. The damaged masonry can then be repaired. The weight also can be transferred to interior walls. Jacking of the ridge and rafters is possible too.

❑ *Cracking due to overloading (or interior movement),* which is fairly uncommon, but may be caused by a point load (often added during an alteration) bearing on a wall of insufficient thickness. If the member has been concealed, such a problem will be difficult to investigate. The addition of interior wall supports or bracing, however, may correct the source of the problem by relieving the load.

❑ *Cracking due to ground tremors from nearby construction, heavy vehicular traffic, or earthquakes,* which is roughly vertical in direction and occurs more toward the center of the building. Buildings exhibiting such cracking should be treated on a case-by-case basis, since serious structural damage has possibly taken place. Consult a structural engineer experienced in such matters.

■ **Bulging of walls.** Masonry walls sometimes show signs of bulging as they age. A wall itself may bulge, or the bulge may only be in the outer withe. Bulging often takes place so slowly that the masonry doesn't crack, and therefore it may go unnoticed over a long period of time. The bulging of the whole wall is usually due to thermal or moisture expansion of the wall's outer surface, or to contraction of the inner withe. This expansion is not completely reversible because once the wall and its associated structural components are "pushed" out of place, they can rarely be completely "pulled" back to their original positions.

The effects of the cyclical expansion of the wall are cumulative, and after many years the wall will show a detectable bulge. Inside the building, separation cracks will occur on the inside face of the wall at floors, walls, and ceilings.

Bulging of only the outer masonry withe is usually due to the same gradual process of thermal or moisture expansion: masonry debris accumulates behind the bulge and prevents the course from returning to its original position. In very old buildings, small wall bulges may result from the decay and collapse of an internal wood lintel or wood-bonding course, which can cause the inner course to settle and the outer course to bulge outward.

When wall bulges occur in solid masonry walls, the walls may be insufficiently tied to the structure or their mortar may have lost its bond strength. Large bulges must be tied back to the structure; the star-shaped anchors on the exterior of masonry walls of many older buildings are examples of such ties (check with local building ordinances on their use). Small bulges in the outer masonry course often can be pinned to the inner course or dismantled and rebuilt.

- Leaning of walls. Masonry walls that lean (invariably outward) represent a serious, but uncommon, condition that is usually caused by poor design and construction practices, particularly inadequate

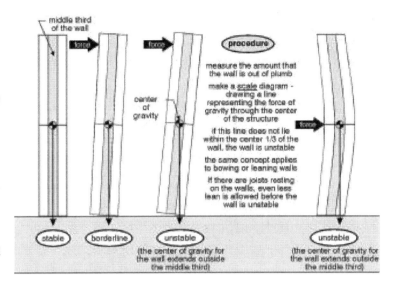

The V3 rule for wall stability

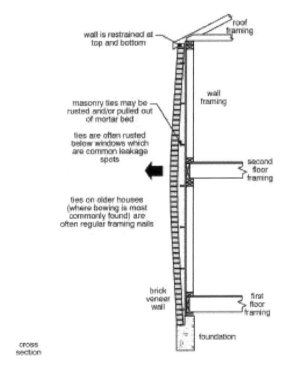

A bowed brick veneer wall

A deteriorated parapet wall that badly needs repointing. Fortunately, the wall has not yet exhibited serious movement, but it will if left unrepaired.

structural tying or poor foundation work. When tilting or leaning occurs, it is often associated with parapets and other upper wall areas, especially those with heavy masonry cornices cantilevered from the wall. Leaning can produce separation cracking on the end walls and cracking on the interior wall face along floors, walls, and ceilings. Leaning walls can sometimes be tied back to the structure and thereby restrained. In such cases, the bearing and connections of interior beams, joists, floors, and roof should be examined.

When large areas or whole walls lean, rebuilding the wall, and possibly the foundation, may be the only answer.

Test: A wall is usually considered unsafe if it leans to such an extent that a plumb line passing through its center of gravity does not fall inside the middle one-third of its base (called the V3 rule). In such an event, consult a structural engineer.

■ **Problems associated with brick veneer walls.** Brick veneer walls are subject to the forces of differential settlement, moisture- and thermal-related cracking, and the effects of freezing and corrosion. Common problems peculiar to brick veneer walls are:

❑ *Cracks caused by wood frame shrinkage,* which are most likely to be found around fixed openings where the independent movement of the veneer wall is restrained. These cracks are also formed early in the life of the building and can be repaired by pointing.

❑ *Bulging,* which is caused by inadequate or deteriorated ties between the brick and the wall to which it is held.

❑ *Vertical cracking at corners or horizontal cracking near the ground caused by thermal movement of the wall,* which is similar to that in solid masonry or masonry cavity walls, but possibly more pronounced in well-insulated buildings because of the reduction in the moderating effect from interior temperatures. Thermal cracks are cyclic and should be filled with a flexible sealant. Where there is severe cracking, expansion joints may have to be installed.

- **Problems associated with parapet walls.** Parapet walls often exhibit signs of distress and deterioration due to their full exposure to the weather, the splashing of water from the roof, differential movement, the lack of restraint by vertical loads or horizontal bracing, and the lack of adequate expansion joints. Typical parapet problems include:

  ❑ *Horizontal cracking at the roof line* due to differential thermal movement between the roof line and the wall below, which is exposed to moderating interior temperatures. The parapet may eventually lose all bond except that due to friction and its own weight and may be pushed out by ice formation on the roof.

  ❑ *Bowing* due to thermal and moisture expansion when the parapet is restrained from lengthwise expansion by end walls or adjacent buildings. The wall will usually bow outward since that is the direction of least resistance.

  ❑ *Overhanging the end walls* when the parapet is not restrained on its ends. The problem is often the most severe when one end is restrained and the other is not.

  ❑ *Random vertical cracking near the center of the wall* due to thermal contraction.

  ❑ *Deterioration of parapet masonry* due to excessive water penetration through inadequate coping or flashing, if any, which when followed by freeze-thaw cycles causes masonry spalling and mortar deterioration.

  Carefully examine all parapet walls. Check their coping and flashing for watertightness and overall integrity. In some cases, structurally unsound parapets can be stabilized and their moisture and thermal movements brought under control by the addition of expansion joints. In other cases, the wall may require extensive repair or rebuilding. All repairs should include adequate expansion joints.

- **Fire damage to brick masonry walls.** Masonry walls exposed to fire will resist damage in proportion to their thickness. Examine the texture and color of the masonry units and probe their mortar. If they are intact and their basic color is unchanged, they can be considered serviceable. If they undergo a color change, consult a qualified structural engineer for further appraisal. Hollow masonry units should be examined for internal cracking, where possible, by cutting into the wall. Such units may need replacement if seriously damaged. Masonry walls plastered on the fire side may have been sufficiently protected and will have suffered few, if any, ill effects. See also Appendix A, Effects of Fire on Structural Systems.

4.6 Chimneys

Chimneys, like parapets, have greater exposure to the weather than most building elements, and have no lateral support from the point where they emerge from the roof. Common problems are:

- **Differential settlement of the chimney** caused by an inadequate foundation. If the chimney is part of an exterior wall, it will tend to lean away from the wall and crack where it is joined to other masonry. In some cases, the chimney can be tied to the building. Consult a structural engineer.

- **Deterioration of masonry near the top** due to a deteriorated cap that allows water into the masonry below and exposes it to freeze-thaw cycles. This cap is often made of a tapered layer of mortar, called a cement wash, that cracks and breaks after several years. Check the cap. If it is mortar and the chimney has a hood, repair the mortar. If it is mortar and the chimney does not have a hood, replace the mortar with a stone or concrete cap. If the cap is stone or concrete, repair it or replace it. Also see Section 2.9.

- **Leaning of the chimney where it projects above the roof** due to deteriorated mortar joints caused either by wind-induced swaying of the chimney or by

A deteriorated chimney cap. Mortar, rather than concrete, is often improperly used (as it was here) to cap the chimney. The masonry below will eventually deteriorate unless the cap is replaced.

sulfate attack from flue gases and particulates within the chimney when the chimney is not protected by a tight flue liner. Deteriorated mortar joints should be pointed, and unstable chimneys or those with a noticeable lean should be dismantled and rebuilt. Chimney-mounted antennas should be removed if they appear to be causing structural distress.

4.7 Wood Structural Components

Wood structural components in small residential buildings are often directly observable only in attics, crawl spaces, or basements. Elsewhere they are concealed by floor, wall, and ceiling materials. Common signs of wood structural problems are sloping or springy floors, wall and ceiling cracks, wall bulges, and sticking doors and windows, although many such problems may be attributable to differential settlement of the foundation or problems with exterior masonry bearing walls (see Sections 4.4 and 4.5).

When failures in wood structural components occur, they usually involve individual wood members and rarely result in the failure of the entire structure. Instead, an elastic adjustment takes place that redistributes stresses to other components in the building. The four types of problems commonly associated with such components in small residential buildings are 1) deflection and warping, 2) fungal and insect attack, 3) fire, and 4) connection failure and improper alteration. Inspect for these problems as follows:

■ **Deflection, warping, and associated problems.** Some deflection of wood structural components or assemblies is common in older buildings and normally can be tolerated, unless it causes loss of bearing or otherwise weakens connections or it opens watertight joints in roofs or other critical locations. Deflection can be arrested by the addition of supplemental supports or strengthening members. Once permanently deflected, however, a wood structural component cannot be straightened.

Warping of individual wood components almost always takes place early in the life of a building. It will usually cause only superficial damage, although connections may be loosened and occasionally there may be a loss of bearing.

Look for the following problems associated with wood structural components:

❑ *Loss of bearing in beams and joists over foundation walls, piers, or columns* due to movements caused by long-term deflection of the wood beams or joists, differential movements of the foundation elements, localized crushing, or wood decay. Check the bearing and connections of all exposed structural elements that are in contact with the foundation and look for

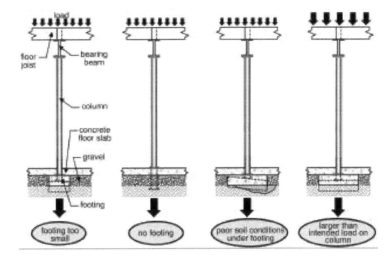

Some reasons for column settling

The sagging floor is caused by settlement of a basement support. It may be possible to jack the floor and its adjacent wall into a level position, but this should be done slowly and carefully.

symptoms of bearing failure where these elements are concealed, such as bowing or sloping in the floor above, and cracking or tilting of foundation walls, piers, and columns.

❑ *Sagging, sloping, or springing of floors* due to foundation settlement, excessive spans, cut or drilled structural elements, overloading, or removal of supporting walls or columns on the floor above or below. Each case must be diagnosed separately. In older buildings, columns or walls that helped support or stabilize the floor above may have been removed during a previous alteration; conversely, partitions, bathrooms, kitchens, or similar remodelings may have been placed on a floor not designed to support such additional loads. Depending on the circumstances, sagging, sloping, or springing floors may be anything from an annoyance to an indication of a potentially serious structural problem. Check below the floor for adequate supports and bearing and for sound connections between structural elements. Look for signs of supporting walls that have been removed, missing joist hangers, and for inappropriate cuts or holes in joists for plumbing, electric, or HVAC lines or ducts. Also look for signs of insect or fungal attack.

Test: Roll a marble or similar heavy round object over suspect floor areas to determine the direction and degree of slope, if any. A carpenter's level also can be used. For large or complex areas, a transit or laser level is more appropriate.

❑ *Floor sagging near stairway openings* due to gradual deflection of the unsupported floor framing. This is a common problem in older houses and usually does not present a structural problem. Correction, if desired, will be difficult since the whole structural assembly surrounding the stair has deformed. Look for signs that a supporting wall below the opening has been removed. Where this has occurred, structural modification or the addition of a supporting column may be required.

❑ *Floor sagging beneath door jambs* resulting from improper support below the jamb. This can be a structural concern. If needed, additional bracing can be added between the joists where the sag occurs, but with some difficulty if above a finished ceiling.

❑ *Cracking in interior walls around openings,* which may be caused by inadequate, deflected, or warped framing around the openings; differential settlement; or on the interior of masonry load-bearing walls, by problems in the exterior masonry wall. Cracking due to framing problems is usually not serious, although it may be a cosmetic concern that can be corrected only by breaking into the wall. See Sections 4.4 and 4.5 for cracking caused by differential settlement or masonry wall problems.

❑ *Sagging in sloped roofs* resulting from too many layers of roofing material, failure of fire retardant plywood roof sheathing, inadequate bracing, or undersized rafters. Sometimes three or more layers of shingles are applied to a roof, greatly increasing its dead load. Or, when an attic story has been made into a habitable space or otherwise altered, collar beams or knee walls may have been removed. A number of factors, such as increases in snow and wind loads, poor structural design, and construction errors result in undersized rafters. Check for all these conditions.

❑ *Failure of Fire Retardant Plywood (FRP)* used at party walls between dwelling units in some townhouses, row houses, and multiple dwellings is not uncommon. Premature failure of the material is due to excessive heat in the attic space. On the exterior, sagging of the roof adjacent to a party wall often is evident. On the interior, check for darkening of the plywood surface, similar to charring, as an indication of failure that requires replacement of the FRP with a product of comparable fire resistance and structural strength.

❑ *Spreading of the roof downward and outward* due to inadequate tying. This is an uncommon but potentially serious structural problem. Look for missing collar beams, inadequate tying of rafters and ceiling joists at the eaves, or inadequate tying of ceiling joists that act as tension members from one side of the roof to the other. Altered trusses can also cause this problem. Check trusses for cut, failed, or removed members, and for fasteners that have failed, been completely removed, or partially disconnected. Spreading can be halted by adequate bracing or tying, but there may be damage to masonry walls below the eaves (see Section 4.5). It is possible that the roof can be jacked back to its proper position. Consult a structural engineer.

❏ *Deflection of flat roofs* due to too great a span, overloading, or improper support of joists beneath the roof. This is a common problem and is usually of no great concern unless it results in leaking and subsequent damage to the structure, or unless it causes water to pond on the roof, thereby creating unacceptable dead loads. In both cases, the roof will have to be strengthened or releveled.

■ **Fungal and insect attack.** The moisture content of properly protected wood structural components in buildings usually does not exceed 10 to 15 percent, which is well below the 25 to 30 percent required to promote decay by the fungi that cause rot or to promote attack by many of the insects that feed on or inhabit wood. **Dry wood will never decay.**

Inspect all structural and non-structural wood components for signs of fungus and insect infestation, including wood stains, fungi, termite shelter tubes, entry or exit holes, signs of tunneling, soft or discolored wood, small piles of sawdust or "frass," and related signs of infestation.

A common location of fungal and insect infestation is where wood door frames touch concrete or earth at grade.

Test: Probe all suspect wood with a sharp instrument and check its moisture content with a moisture meter. Wood with a meter reading of more than 20 to 25 percent should be thoroughly examined for rot or infestation. Sound wood will separate in long fibrous splinters, but decayed wood will lift up in short irregular pieces. See Appendix B, Wood Inhabiting Organisms, for more detailed information.

Exterior building areas or components that should be checked are:

❏ *Places where wood is in contact with the ground,* such as wood pilings, porch and deck supports, porch lattices, wood steps, adjacent fences, and nearby wood piles.

- ❏ *Foundation walls* that might harbor shelter tubes, including tubes in the cracks on wall surfaces.
- ❏ *Frames and sills* around basement or lower level window and door frames, and the base of frames around garage doors.
- ❏ *Wood framing* adjacent to slab-on-grade porches or patios.
- ❏ *Wood near or in contact with roofs,* drains, window wells, or other areas exposed to periodic wetting from rain or lawn sprinklers, etc.

Interior areas or components to be checked for rot or infestation are:

- ❏ *Spaces around or within interior foundation walls* and floors, crawl spaces, piers, columns, or pipes that might harbor shelter tubes, including cavities or cracks.
- ❏ *The sill plate* that covers the foundation wall, and joists, beams, and other wood components in contact with it.
- ❏ *Wood frame basement partitions.*
- ❏ *Baseboard trim* in slab-on-grade buildings.
- ❏ *Subflooring* and joists below kitchen, bathroom, and laundry areas.
- ❏ *Roof sheathing* and framing in the attic around chimneys, vents, and other openings.

Damage to wood from fungal or insect attack usually can be repaired at a reasonable cost by replacing or adding supplemental support to affected components after the source of the problem has been corrected. Damage is rarely severe enough to seriously affect the structural stability of a building, although individual members may be badly deteriorated. Consult an exterminator when evidence of insect attack is found.

■ **Fire-damaged wood.** When exposed to fire, wood first "browns," then "blackens," then ignites and begins to char at a steady rate. The charred portion of the wood loses its structural strength, but the clear wood beneath does not, unless it undergoes prolonged heating.

The remaining strength of wood exposed to fire can be determined by removing the char and estimating the size and strength of the new cross section. Damaged structural members may be reinforced by bolting additional structural members in a configuration that restores their original design strength. Consult a structural engineer before repairing major structural beams or girders. See also Appendix A, Effects of Fire on Structural Systems.

■ **Connection failure and improper alteration.** Open-web wood trusses are commonly used in small residential buildings as roof and floor structures. Those trusses with wood chords and wood webs usually use metal plate connectors. Where there is evidence of moisture, examine connectors for corrosion and for loss of embedment due to distortion of truss members. Also check glue laminated timber beams for delamination where there is evidence of moisture. Where possible, examine truss, rafter, and joist hangers for corrosion and proper nailing.

Buildings are often altered incrementally by the addition of pipes and ducts when unfinished spaces, such as basements and attics, are made habitable or when kitchens and bathrooms are remodeled. Observe joists, rafters, and beams for holes cut through them, especially for any size cutouts at or near their top or bottom. Observe wood trusses for cuts through chords and webs of open web trusses and webs and flanges of plywood trusses.

4.8 Iron and Steel Structural Components

Metal structural components used in small residential buildings are usually limited to beams and pipe columns in basements, angles over small masonry openings, and beams over long spans elsewhere in the structure. These components are almost always made of steel, although in buildings erected before 1890 to 1900 they may be of cast or wrought iron. While cast iron is weaker in tension than steel, when found in small buildings it is rarely of insufficient strength unless it is deteriorated or damaged.

Problems with iron and steel structural components usually

This fire-damaged wood should be carefully probed to determine the extent of charring. In this case, the wood was replaced.

center on corrosion. Inspect them as follows:

- **Lintels and other embedded metal components in exterior masonry walls** can corrode and in time become severely weakened themselves. Rain and snow often contain carbonic, sulfuric, nitric, or hydrochloric acid that lowers the pH of rain water, thereby accelerating corrosion. Check all embedded iron and steel to determine its condition.

   Make sure lintels have adequate bearing. Corrosion can also displace surrounding masonry; see Section 4.5.

- **Columns** should be checked for adequate connections at

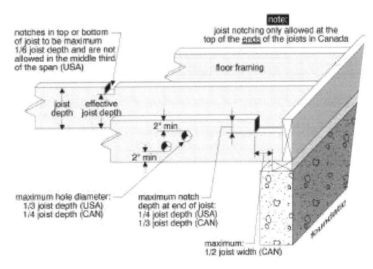

Joist notching and drilling criteria

Failure of a concrete lintel and sill due to differential settlement of the building. Permanent repairs will be quite expensive. Note the makeshift shoring of the lintel.

their base and top, and for corrosion at their base if they rest at ground level. Eccentric (off-center) loading or noticeable tilting of columns should be remedied.

■ Beams should be checked for bearing, adequate connections to the structure, and deflection. Bearing can be significantly reduced on pilasters, piers, or columns in differentially settled buildings; inspect such conditions carefully (see Section 4.4). Beams in small residential buildings rarely deflect. If deflection is found, however, the cause should be determined and supplemental supports or plates should be added to correct the problem.

■ Fire damage to iron and steel structural components should be carefully inspected. Iron and steel rapidly lose their load-bearing capacity when exposed to fire and will undergo considerable expansion and distortion. In general, a structural iron or steel member that remains in place with negligible or minor distortions to its web, flanges, or end connections should be considered serviceable. Sagging or bent members, or those with a loss in bearing capacity should be replaced or reinforced with supplemental plates.

Test: When the quality or composition of an iron or steel structural component is in doubt, a small sample of the metal (called a "coupon") may be removed from a structurally unimportant location and sent to a testing laboratory for evaluation. The sample should be tested in accordance with

ASTM E8, *Standard Test Methods for Tension Testing of Metallic Materials*, and ASTM E9, *Standard Test Methods of Compression Testing of Metallic Materials at Room Temperature*. Such work should be performed under the auspices of a structural engineer.

4.9 Concrete Structural Components

Concrete is commonly used for grade and below-grade level floors and for footings. It also may be used for foundations, beams, floors above grade, porches or patios built on grade, exterior stairs and stoops, sills, and occasionally as a precast or poured-in-place lintel or beam over masonry openings. Concrete structural components are reinforced. Welded steel wire mesh is used in floors at and below grade, patios built on grade, walks and drives, and short-span, light-load lintels. All other concrete structural components usually are reinforced with steel bars.

Inspect for the following:

- **Cracking at corners or openings in concrete foundations below masonry exterior walls** due to drying shrinkage of concrete walls that are prevented from contracting by the mass of the masonry above. This cracking will occur early in the life of the building. Minor cracks can be filled with mortar and major cracks with concrete epoxy grout.

- **Cracking of interior slabs on grade** is usually due to shrinkage or minor settlement below the slab. If cracking is near and parallel to foundation walls, it may have been caused by the movement of the walls or footers (see Section 4.4). Cracking can also result from soil swelling beneath the slab, a condition that may be caused by water from clogged or broken basement or footer drains. Rarely is such cracking structurally harmful to the building.

- **Cracking of exterior concrete elements,** such as porches, patios, and stairs, is usually due to heaving from frost or nearby tree roots, freeze-thaw cycles, settlement, or a combination of these conditions. It is compounded by the use of deicing salts. Such cracking rarely presents a structural problem to the building, but is often a practical problem that can best be remedied by replacing the concrete and providing the new work with more stable support. Cracks in existing concrete elements that are not seriously deteriorated may be cyclical and can be filled with a flexible sealant.

- **Fire damage to concrete structural components** should be thoroughly evaluated. Concrete heated in a building fire will lose some compressive strength, although when its temperature does not exceed 550 °F (290 °C) most of its strength eventually will be recovered. If the concrete surface is intact, it can usually be assumed to be in adequate condition. Superficial cracking can be ignored. Major cracks that could influence structural behavior are generally obvious and should be treated on a case-by-case basis. Cracks can be sealed by injecting concrete epoxy grout. Paints are available to restore the appearance of finely cracked or crazed concrete surfaces. See also Appendix A, Effects of Fire on Structural Systems.

Tests: Two specialized tests may sometimes be useful for estimating the quality, uniformity, and compressive strength of in-situ concrete. The first is the Windsor Probe, a device that fires a hardened steel probe into concrete. See ASTM C803, *Standard Test Method for Penetration Resistance of Hardened Concrete*. The second test is the Schmidt Hammer, which measures the rebound of a hardened steel hammer dropped on concrete. See ASTM C805, *Standard Test Method for Rebound Number of Hardened Concrete*.

For additional information about inspecting and repairing concrete, consult the following publications from the American Concrete Institute: ACI 201.1R, *Guide for Making a Condition Survey of Concrete in Service*; ACI 364.1R, *Guide for Evaluation of Concrete Structures Prior to Rehabilitation*; and ACI 546.1R, *Concrete Repair Guide*.

5
Electrical System

Electrical systems for small residential buildings are usually simple in concept and layout. Primary components are the service entry, panelboard, and branch circuits. In unaltered buildings built since about 1940, the electrical system is likely to be intact and safe, although it may not provide the capacity required for the planned reuse of the building. Electrical capacity can be easily increased by bringing additional capacity in from the street and adding a larger panelboard between the service entry and the existing panel. Existing circuits can continue to use the existing panel and new circuits can be fed through the new panel.

The electrical systems of small residential buildings built prior to about 1940 may require overhaul or replacement, depending on rehabilitation plans and the condition of the electrical system. Parts of these older systems may function very adequately and they can often be retained if the rehabilitation is not extensive and the load-carrying capacity is adequate.

A thorough and informed assessment of the electrical system will determine the extent to which it can be reused. This assessment should be conducted only by a qualified electrician who is experienced in residential electrical work.

When universal design is a part of a rehabilitation, consult HUD publication *Residential Remodeling and Universal Design* for detailed information about electrical devices.

Assess the capacity of the building's existing electrical service in accordance with Figure 5.1.

The safety standards for the following assessment procedures are generally based on the requirements of the *National Electrical Code*.

5.1
Service Entry

Inspect for the following conditions in the electrical service between the street and the main panelboard:

- **Overhead wires.** Check that overhead wires from the street are no lower than 10 feet above the ground, not in contact with tree branches or other obstacles, and not reachable from nearby windows or other accessible areas. Make sure that the wires are securely attached to the building with insulated anchors, and have drip loops where they enter the weatherhead. Spliced connections at the service entrance should be well wrapped, and bare wires from the street should be replaced by the utility company. Wires should not be located over swimming pools.

- **Electric meter.** Check that the electric meter and its base are weatherproof, and that the meter is functional, has not been tampered with, and is securely fastened. Advise the utility company of any problems with the meter.

- **Seismic vulnerability.** If the building is in a seismic zone, check the electrical service for vulnerability to differential movement between the exterior and interior. Look for flexible connections.

- **Service entrance conductor.** Ensure that the service entrance conductor has no splices and that its insulation is completely intact. If the main panelboard is located inside the building, the conductor's passage through the wall should be sealed against moisture. Where aluminum conductors are used, their terminations at all service equipment should be cleaned with an oxide inhibitor and tightened by an electrician or replaced with equal capacity copper conductors. When it is necessary to replace an overhead service entry, have it replaced with an underground service entry.

- **Type of power available.** Not every jurisdiction provides the same kind of electrical power. Philadelphia, for example, has two-phase electrical power in some locations rather than the more common single-phase. Check with the power company to determine the characteristics of the power available.

5.2
Main Panelboard (Service Equipment)

The main panelboard is the distribution center for electric service within the building and protects the house wiring from overloads. Inspect the panelboard as follows:

- **Condition and location.** Check the overall condition of the panelboard. Water marks or rust on a panel mounted inside the building may indicate water infiltration along the path of the service entrance conductor. Panelboards mounted outdoors should be watertight and tamper proof. Panels mounted indoors should be located as closely as possible to where the service entrance conductor enters the building and should be easily accessible. The panelboard should have a workable and secure cover.

- **Amperage rating.** The amperage rating of the main disconnect should not be higher than the amperage capacity of the service entrance conductor or the panelboard. If the rating is higher (indicating unapproved work has been done), more branch circuits may be connected to it than the service entrance conductor is capable of supplying. This is a serious hazard and should be corrected.

- **Voltage rating.** The voltage rating of the panelboard (as marked on the manufacturer's data plate) should match the voltage of the incoming electrical service.

Figure 5.1
Assessing Electrical Service Capacity (Ampacity)

To determine the capacity (measured in amperes) of the building's existing electrical service at the main panelboard, check the following:

- The ampacity of the service entry conductor, which may be determined by noting the markings (if any) on the conductor cable and finding its rated ampacity in the *National Electrical Code*, Table 310-16, or applicable local code. If the service entry conductor is in conduit, look for markings on the conductor wires as they emerge from the conduit into the panelboard. If all conductors are unmarked, have an electrician evaluate them.

- The ampere rating on the panelboard or service disconnect switch, as listed on the manufacturer's data plate.

- The ampere rating marked on the main circuit breaker or main building fuse(s). This rating should never be higher than the above two ratings; if it is, the system should not be used until it is evaluated by an electrician.

The building's service capacity is the lowest of the above three figures. Once the service ampacity has been determined, compare it to the estimated ampacity the building will require after rehabilitation. If the estimated ampacity exceeds the existing ampacity, the building's electrical service will need upgrading. The method for estimating required ampacity is found in the *National Electrical Code*, Article 220.

Similarly, the service capacity of each branch circuit can be determined by checking the markings on each branch circuit conductor. If no markings can be found, a plastic wire gauge may be used to measure the wire size (with the power disconnected), although an experienced person can often determine the size by eye. Find the ampere rating of the conductor, either by its markings or wire size, in the *National Electrical Code*, Table 310-16, or applicable local code.

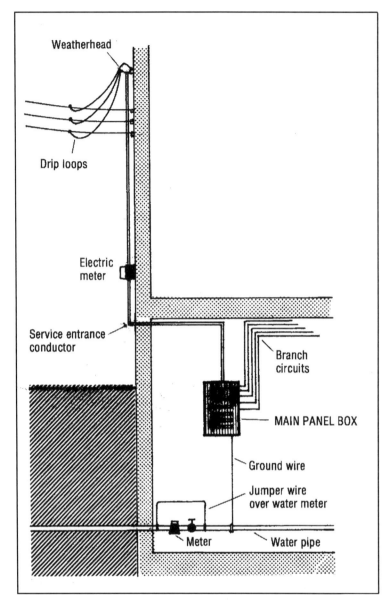

Typical electrical service entry and main panelboard for a single family residence. This type of grounding applies only if the water pipe is metal. If the water pipe is plastic, a separate driven ground rod is required.

Test: The actual voltage rating of the incoming electrical service can be checked with a voltmeter. This test should be performed by an electrician. Usually three service conductors indicates 120/240 volt current, and two conductors indicates 120 volt current.

- **Grounding.** Verify that the panelboard is properly grounded. Its grounding conductor should run to an exterior grounding electrode or be clamped to the metal water service inlet pipe between the exterior wall and the water meter. If it is attached on the house side of the meter, the meter should be jumpered to ensure proper electrical continuity to the earth. Make sure that the ground conductor is securely and properly clamped to the pipe—often it is not, and occasionally it is disconnected altogether. Ensure also that the grounding conductor is not attached to a natural gas pipe, to an inactive pipe that may be cut off on the exterior side of the wall, or to a pipe that is connected to a plastic water service entry line. If the grounding conductor is attached to an exterior grounding electrode driven into the earth, verify that the electrode is installed in accordance with local code. Many older buildings will have the ground connected to the cold water pipe. If this is the case and the building needs to conform to the current code, an alternate ground is required.

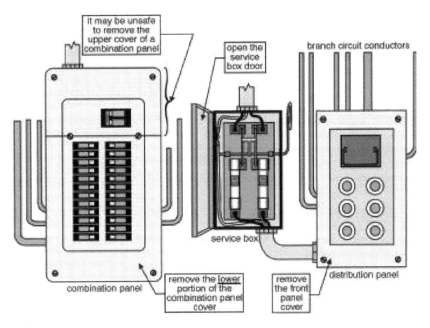

Typical service equipment

Test: An electrical ground (resistance-to-ground) test may be used to determine whether the electrical system is well grounded to the earth. The test requires the use of an ohmmeter and should be performed by an electrician.

- **Overcurrent protection.** Check the rating of the fuse or circuit breaker for each branch circuit. The amperage of the fuse or circuit breaker should not exceed the capacity of the wiring in the branch circuit it protects. Most household circuits use #14 copper wire, which should have 15 amp protection. There may be one or more circuits with #12 copper wire, which should have 20 amp protection. Large appliances, such as electric water heaters and central air conditioners, may require 30 amp service, which is normally supplied by #10 copper wire. If there is an electric range, it would require a 40 or 50 amp service with #6 copper wire. Central air conditioning equipment will have an overcurrent protection requirement on the nameplate. Aluminum wire must be one size larger than copper wire in each case (e.g., #14 to #12), but it should not be used for 15 and 20 amp circuits. See Figure 5.1 for determining wire size.

Make sure that no circuit has a fuse or circuit breaker with a higher ampere rating than its wiring is designed to carry. Air conditioners and other equipment with motors may have circuit breakers up to 175 percent ampacity of the conductor rating to allow for starting current. Look near the panelboard for an inordinate number of new or blown fuses, or breakers taped in the "on" position. Be suspicious of 20 or 25 amp fuses on household lighting circuits. These are signs of frequent overloads and inadequate electrical service. Other indications of overloading are the odor of burned insulation, evidence of melted insulation, discolored copper contact points in the fuse holders, and warm fuses or circuit breakers.

Test: Flip all circuit breakers on and off manually to make sure they are in good operating condition. A commercially available circuit breaker and resistance tester, which can simulate an overload condition, can be used to test each breaker. Such a test should be performed by an electrician. Note

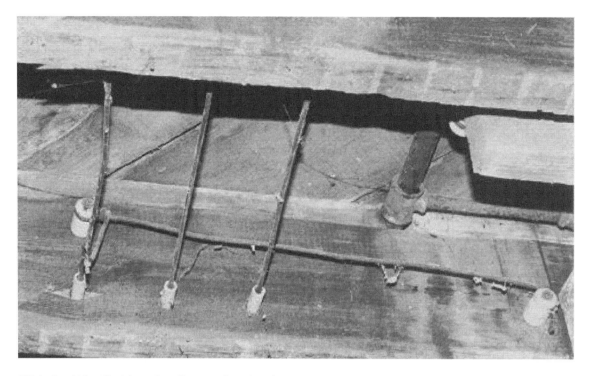

This knob and tube wiring is in good condition except for a piece of broken insulation (top of photograph).

that this test is not recommended for computers, VCRs, clocks, and many similar devices.

Many older residential buildings have more than one panelboard or fused devices. Check that all supplementary overcurrent devices are located in metal boxes and that they are not in the vicinity of easily ignitable materials. All panelboards must have covers. It should be possible to turn off all electrical power to a dwelling from a single location.

5.3 Branch Circuits

The oldest types of residential wiring systems are seldom encountered today. They include open wires on metal cleats, wiring laid directly in plaster, and wiring in wooden molding. These systems proved quite hazardous. The oldest wiring system that may still be acceptable, and one still found fairly often in houses built before 1930, is "knob and tube." This system utilizes porcelain insulators (knobs) for running wires through unobstructed spaces, and porcelain tubes for running wires through building components such as studs and joists. Note whether knob and tube wiring splices are mechanically twisted, soldered, and taped, as required. Knob and tube wiring should be replaced during rehabilitation; but if it is properly installed, needs no modification, has adequate capacity, is properly grounded, has no failed insulation, and is otherwise in good condition, it can be an acceptable wiring system and is still legal in many localities. Check with local building code officials. Also check the terms and conditions of the home insurance policy in force to see if knob and tube wiring is excluded. The greatest problem with such wiring is its insulation, which turns dry and brittle with age and often falls off on contact, leaving the wire exposed.

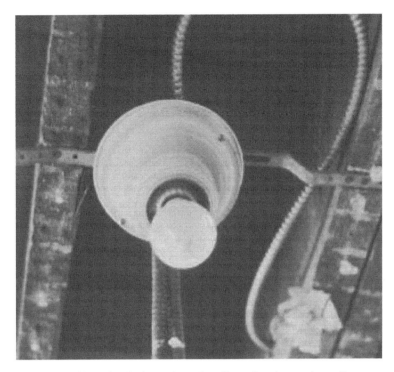

The armored cable and junction box are in good condition and can be reused, even if the lighting fixture is relocated.

Insulation that can be seen to have failed also will likely have failed where wiring is concealed. If any failed insulation is observed, the knob and tube wiring should be replaced.

Other approved wire types include:

- NM (non-metallic) cable, often called by the trade name "Romex," a plastic covered-cable for use in dry locations (older NM cable may be cloth covered).
- NMC, similar to NM but rated for damp locations.
- UF (underground feeder), a plastic-covered waterproof cable for use underground.
- AC (armored cable), also called BX, a flexible metal-covered cable.
- MC (metal-clad cable), a flexible metal-covered cable with a green insulated ground conductor.
- EMT (electrical metallic tubing), also called "thinwall," a metal conduit through which the wires are run in areas where maximum protection is required.

Check branch circuits for the following:

- **Marking.** The function of each branch circuit should be clearly and legibly marked at its disconnect, fuse, circuit breaker, or on the directory on the panelboard.
- **Connected loads.** Trace branch circuit conductors to determine that their connected load does not exceed their rating (e.g., a 30 amp clothes dryer connected to a 20 amp circuit). Generally speaking, each dwelling unit should have two to four 15 amp circuits for lighting and convenience outlets; two 20 amp circuits for appliances in the kitchen, dining, and laundry areas; and separate circuits of appropriate ampacity for large appliances such as dryers, ranges, disposals, dishwashers, and water heaters. See Section 3.4 for additional kitchen electrical service information. Check the size and length of all branch circuit wiring against the requirements of the local electrical code. Buildings built before 1980 may be considered to have an inadequate number of circuits because present day codes require a separate laundry circuit and a separate circuit for the bathroom receptacle. For air conditioning units, many local codes will allow one wire size smaller than called for in the disconnect.

Test: A voltmeter may be used to measure voltage drop due to excessive branch circuit length, poor wiring connections, or undersized wire. Measurements must be made under a connected load. This test should be performed by an electrician.

- **Grounding.** It is best that all circuits be grounded to the panelboard, but this was not required by the *National*

Electrical Code prior to 1965. Do not assume that circuits in metal cable are grounded without testing each outlet. Also, do not assume that three-prong plug convenience outlets are connected to ground. Remove each one to observe the presence of a connected ground wire. Check to see whether GFI (ground fault interruption) type receptacles have been installed in laundries, kitchens, and bathrooms, and test their operation. These types of receptacles were not required before 1990, but are easily installed as replacements.

Test: Commercially available circuit analyzers can be used for checking the following circuit conditions: open ground, open hot, open neutral, hot/ground reversed, hot/neutral reversed. Operation of these analyzers varies by manufacturer.

■ **Condition and safety.** Check that all wire types and equipment are installed properly in accordance with good practice. Check the conductors' exposure to possible damage or abrasion. Look for proper fastening, clearance, and frayed or damaged insulation. Make certain that all wire splices are made in work boxes and that all boxes for splices and switches have cover plates. Check all exterior receptacles to make sure they are of the waterproof type.

Test: A megohm test may be used for detecting deteriorated insulation. It requires a Megger tester and operates at high voltage. With the electrical service disconnected, branch circuits should read at least one megohm to ground. If lights or appliances are connected to the circuit, readings should be at least 500,000 ohms. This test should be performed by an electrician. A visual inspection of insulation on accessible circuits will usually determine whether additional tests should be performed by an electrician.

Look for unprotected wire runs through ducts and other inappropriate areas. Inspect for evidence of "handyman tampering" (e.g., unconventional splices), and if found in one location, expect it to be more widespread. Check for surface-mounted lamp cord extension wiring. It is dangerous and must be removed. It is best to remove all unused wiring or wiring that will be abandoned during rehabilitation work to avoid future confusion or misuse.

■ **Aluminum wire.** Aluminum wire was used in residential buildings primarily during the 1960s and early 1970s. Inspect with local code requirements in mind. Be sure that aluminum wire is attached only to approved devices (marked "CO-ALR" or "ICU-AL") or to approved connectors. Problems with aluminum wiring occur at connections, so feel all cover plates for heat, smell for a distinctive odor in the vicinity of outlets and switches, and look for sparks and arcing in switches or outlets and for flickering lights. Also check for the presence of an oxide inhibitor on all aluminum wire connections. All such conditions should be corrected. Aluminum wire should not be used on 15 and 20 amp circuits. Whenever possible, aluminum wire and its devices should be replaced with copper wire and devices appropriate for copper. If aluminum wiring is not replaced, it must be frequently inspected and maintained.

■ **Smoke Detectors.** Check to see if buildings have functioning smoke detectors. Detectors should be wired to a power source, and also should contain a battery. Most likely, buildings built before 1970 will not have detectors, but they should be added.

6
Plumbing System

A thorough assessment of the plumbing system will determine the extent to which it, like the electrical system, can be reused. Older piping in particular may require replacement, but other parts of the system may function very adequately and can be retained if rehabilitation is not extensive. Also see Sections 3.3 and 3.4 for additional plumbing information.

If the plumbing system appears to be functioning properly after checking, consider the effects of additional loads that may be imposed on the system by any rehabilitation that might be planned for the building.

Assess the capacity of the building's existing plumbing system in accordance with Figure 6.1.

6.1
Water Service Entry

Inspect the following water service components:

- **Curb valve** (also known as the curb cock or curb stop). The curb valve is located at the junction of the public water main and the house service main, usually near the street. Locate it and check its accessibility and condition. The curb valve is usually the responsibility of the municipal water department.

- **House service main.** The house service main begins at the curb valve and ends at the inside wall of the building at the master shutoff valve. The main is normally laid in a straight run between the two and its location can thereby be traced. Codes require that the main be at least 10 feet (3 m) away from the sanitary sewer or located on a plane one foot above it. Mains made of galvanized steel last about 20 to 30 years under normal soil and water conditions, although joints may leak sooner. If the building is approaching or more than 40 years old, consider replacing the main.

  Test: Leaks in the main can be detected by inspecting for unexplained sources of ground water over the path of the main or by listening with a stethoscope for underground water flow. Stop all water flow inside the building before using this device. If the water meter is located near the street, leaks in the service main can also be detected by turning off all water in the building and watching the meter to see whether it continues to register a water flow.

  Check the main where it enters the building; older lines are sometimes made of lead.

  Test: If a lead main is found, have the water analyzed and replace the piping if lead content exceeds 50 parts per billion (0.05 ppm).

- **Master shutoff valve.** A master shutoff valve should be located where the house service main enters the building. If the water meter is located inside the building, look for another water meter outside of the building and two shutoff valves, one on the street side and one on the house side. If a valve appears corroded or damaged, have a plumber check to see that it is operable and not frozen into the open position. The shutoff valve should include a bleed valve for draining the building's interior distribution piping.

- **Water meter.** The water meter is normally the property of the municipal water company and may be located near the street, adjacent to the house, or within the house. Check the meter connections and supports, and inspect for adjacent plumbing constrictions that may reduce the building's water pressure. If the water meter is located inside the house, look for two shutoff valves, one on the street side and one on the house side of the meter.

- **Seismic vulnerability.** If the building is in seismic zones 3 or 4 (California and portions of Alaska, Arkansas, Hawaii, Idaho, Missouri, Montana, Nevada, Oregon, Utah, Wyoming, and Washington), check the water service for vulnerability to differential movement where the piping enters the building. Look for adequate clearance.

Figure 6.1
Assessing Water Supply Capacity

The minimum size of the water service entry should be approximately as follows:

Number of dwelling units served	Size of galvanized steel pipe (NPS/DN)	Size of type K copper pipe (NPS/DN)
1	1 to 1-1/4 inch (25 to 32 mm)	3/4 to 1 inch (20 to 25 mm)
2	1-1/4 to 1-1/2 inch (32 to 40 mm)	1 to 1-1/4 inch (25 to 32 mm)
3, 4	1-1/2 to 2 inch (40 to 50 mm)	1 to 1-1/4 inch (25 to 32 mm)

The following minimums should generally apply to fixture supply pipes within the building (these are code minimums; 15 psi [103 kPa] at each fixture is preferable to the lower figures listed here):

	Minimum pipe size (NPS/DN)	Minimum flow rate	Minimum flow pressure
Kitchen sink	1/2" (15 mm)	2.5 gpm (9.5 L/min)	8 psi (55 kPa)
Lavatory	3/8" (10 mm)	2.0 gpm (7.5 L/min)	8 psi (55 kPa)
Shower	1/2" (15 mm)	3.0 gpm (11.5 L/min)	8 psi (55 kPa)
Bathtub	1/2" (15 mm)	4.0 gpm (15 L/min)	8 psi (55 kPa)
Toilet	3/8" (10 mm)	3.0 gpm (11.5 L/min)	8 psi (55 kPa)
Dishwasher	1/2" (15 mm)	2.75 gpm (10.5 L/min)	8 psi (55 kPa)
Laundry	1/2" (15 mm)	4.0 gpm (15 L/min)	8 psi (55 kPa)
Hose bib	1/2" (15 mm)	2.5 gpm (9.5 L/min)	8 psi (55 kPa)

Test: The capacity of the interior water distribution piping should be checked by running the water in all the fixtures in a fixture group (such as a bathroom) and observing whether water flow is adequate. Water flow can be more precisely tested at each fixture by using a water pressure gauge to determine the fixture's pressure in psi, or by clocking the time it takes to fill a gallon jug from each fixture (e.g., a kitchen sink faucet should be able to fill a gallon jug in 24 seconds; that is, 60 seconds divided by 2.5 gpm).

If the service entry is correctly sized but water flow is low throughout the building, the problem will either be the jurisdiction's water pressure or it will be somewhere between the building inlet and the first fixtures on the line. Check for external restrictions in the supply main, such as undersized piping (particularly around the water meter), partially closed valves, or kinks in the piping. If there are no apparent external restrictions, the piping is probably clogged by rust and/or mineral deposits and should be replaced. If water flow is low in only one set of fixtures, examine the fixture risers in a like manner and inspect plumbing fixtures for flow restrictors, clogged aerators, or malfunctioning faucets. If water flow is lower in hot water faucets than cold, suspect problems with the water heater or, more likely, buildup of rust and mineral deposits in the supply lines, since this buildup occurs faster in hot water.

If new fixtures are to be added to the distribution system, have a plumber determine whether the existing piping can carry the additional load by checking the size and condition of the piping and calculating the water demands of the fixture(s) to be added.

6.2 Interior Water Distribution

All piping, regardless of composition, should be checked for wet spots, discoloration, pitting, mineral deposits, and leaking or deteriorated fittings. Fixture risers tend to remain in better condition than supply mains, so their inspection is not as critical. The water flow from all fixtures should be checked.

Test: To check water flow, run several plumbing fixtures at once and observe the flow rate at the smallest spout. Rusty water indicates that galvanized steel is present somewhere in the line or, if it only appears from the hot water side, that there is rust in the water heater.

Inspect the following water distribution components:

- **Distribution piping.** Distribution piping consists of supply mains and fixture risers.

Most supply mains can be inspected from the basement or from crawl spaces, but the fixture risers are usually concealed within walls and cannot be readily examined. The two most important factors in assessing distribution piping are the piping's material and age.

Test: Pressure test any piping suspected of having leaks.

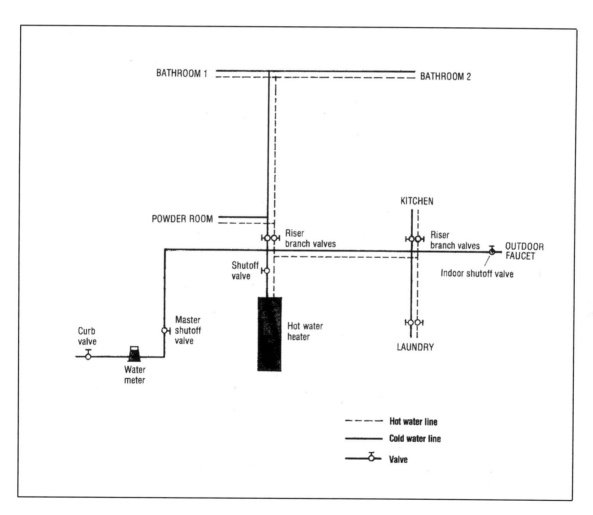

Typical water distribution system schematic for a single-family residence

Galvanized pipe sections removed from an older house. Mineral deposits and corrosion within the pipe had severely reduced water flow to the plumbing fixtures.

❑ *Galvanized steel piping* is subject to rusting and accumulating more mineral deposits than most other piping materials. Depending on the quality of the pipe and its joints and the mineral content of the water it carries, the service life of galvanized steel piping is anywhere from 20 to 50 years. Rusted fittings and rust-colored water, particularly from hot water lines, are signs of advanced deterioration. Low rates of flow and low water pressure are likely to be caused by galvanized steel piping clogged with rust and mineral deposits. During rehabilitation, if galvanized steel piping is exposed, consider replacing it.

❑ *Brass piping* is of two varieties, yellow and red. Red is more common and has the longer service life—up to 70 or more years. The service life of yellow brass is about 40 years. Old brass piping is subject to pinhole leaking due to pitting caused by the chemical removal of its zinc content by minerals in the water. Often, water leaking from the pinhole openings will evaporate before dripping and leave whitish mineral deposits. Whitish deposits may also form around threaded joints, usually the most vulnerable part of a brass piping system. Brass piping with such signs of deterioration should be replaced.

❑ *Copper piping* came into widespread use in most parts of the country in the 1930s and has a normal service life of 50 or more years. Copper lines and joints are highly durable and usually not subject to clogging by mineral deposits. Such piping need not be replaced unless there are obvious signs of deterioration, leakage, or restriction of water flow. Leakage usually occurs near joints and at supports.

❏ *Plastic piping* (ABS, PE, PB, PVC, and CPVC) is a relatively new plumbing material and, if properly installed, supported, and protected from sunlight and mechanical damage, should last indefinitely. However, there are several class action lawsuits pending at this time concerning polybutylene pipe and fittings used inside and outside buildings. Funds resulting from these suits are controlled by local jurisdictions. Check with local authorities or consumer advocate groups for details. Some codes restrict the use of plastic piping. Consult the local building official.

Some newer buildings use a manifold off the water main to distribute cold water and a manifold off the water heater to distribute hot water. From the manifold, flexible plastic pipes are snaked through floors and walls to each plumbing fixture. Check manifolds closely for signs of corrosion and leaks. When testing water flow, all the fixtures off a manifold should be run at the same time.

❏ *Lead piping* may be found in very old structures and may pose a health hazard to building occupants.

Test: If lead piping is found, have the water analyzed for lead content and replace the piping if lead content exceeds 50 parts per billion (0.05 ppm).

❏ *Mixed metal piping* that is a mixture of galvanized steel and copper or brass piping is a sign of potential trouble and should be closely inspected for corrosion due to galvanic action. Where pipes of dissimilar metals are connected, be certain a dielectric coupling separates them. Also, metal pipes and dissimilar metal supports need to be separated to avoid corrosion. Check connections to plumbing and HVAC equipment, such as water heaters and boilers, to be certain that pipes and connections are the same metal or, if of different metals, that a dielectric coupling separates them. No separation is needed between metal and plastic pipe.

■ **Thermal protection.** Examine all water distribution lines for exposure to freezing conditions and look for signs of previous water damage from burst joints or piping. Determine whether the piping remains exposed to freezing and whether any planned rehabilitation work will block the moderating effects of the building's interior temperature from any part of the piping. Consider the costs and benefits of insulating hot water lines during the building's rehabilitation.

6.3 Drain, Waste, and Vent Piping

Determine drain, waste, and vent (DWV) capacity as described in Figure 6.2. Inspect the DWV piping as follows:

■ **Fixture traps.** Fixture traps are generally U-shaped and designed to hold a water seal that blocks the entry of sewer gasses through the fixture drain. Check all fixture drains for evidence of water seal loss; such drains usually emit the odor of sewer gas. The water seal in water closets can be visually verified, while other fixtures can be checked with a dipstick.

Test: Refill any empty traps and discharge their fixtures to determine whether the water seal was lost due to a plumbing malfunction or through evaporation due to lack of use. If, after operation, the traps are again "pulled," the problem is caused either by self-siphonage because of improper plumbing design, obstructions in the venting system, or the lack of a vent. In this case, first check for the presence of S-traps under the fixture, which may cause the self-siphonage and are no longer allowed by plumbing codes. If S-traps are not present, thoroughly check the venting system. Even if all plumbing fixtures are to be replaced, this assessment process should be performed to reveal problems in the overall system.

■ **Vents.** Vents equalize the atmospheric pressure within the waste drainage system to prevent siphoning or "blowing" of the water seals in the building's fixture traps. Vents

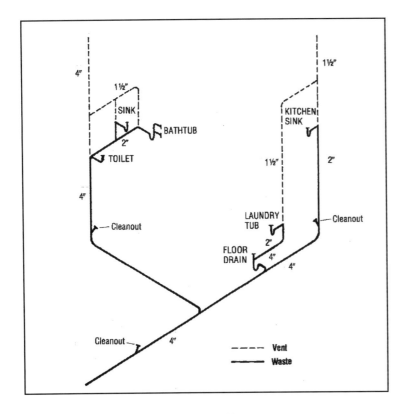

Typical DWV piping schematic for a single-family residence

should be unobstructed and open high enough above the roof to prevent snow closure. Vents that terminate outside an exterior wall or terminate near a building opening (such as a dormer window) are prohibited by building codes, although under certain conditions they may be acceptable. Check vent lines for damage caused by building movement or settlement or by the sagging of individual building components.

Test: Discharge several fixtures simultaneously while observing fixture traps; water movement greater than one inch in the trap indicates inadequate or obstructed venting that must be corrected. Also, fill a sink, lavatory, or tub with water and listen to the fixture drain. If a gurgling sound is heard, it usually indicates a venting problem.

An S-trap that can cause self-siphonage and loss of the water seal. All such traps should be replaced.

Figure 6.2 Assessing DWV Capacity

The installed capacity of an existing DWV system can be estimated by measuring the size of each DWV stack and, using the local plumbing code, finding the allowable number of fixtures that can drain into it. This is a relatively simple process.

- **Drain lines.** Drain lines direct waste water from the fixture trap through the building to the sewer. Because the waste drainage system operates by gravity, drain lines must be of adequate size and slope to function properly. Minimum slope should be 1/8 inch per foot (1:00) Cleanouts should be located near the juncture of all main vertical drain pipes that enter the building drain. Check for low spots on long horizontal runs caused by inadequate support, and for damage or distress caused by building settlement or movement. Check also for drains with pipes of dissimilar metals that are not separated by a dielectric coupling to prevent corrosion. Metal pipes with dissimilar metal supports need to be separated to avoid corrosion.

Test: Test the waste drainage system by discharging several fixtures simultaneously. Look for "boiling" or back-up in the lowest fixture in the building. This indicates a clogged or malfunctioning main building drain between the building and the public sewer. Most often such a problem is caused by tree roots that have clogged the line.

Test: Oil of peppermint or smoke can be used to check the hydraulic integrity of DWV systems by inserting either substance in the system (the oil through a roof vent, the smoke through a trap) and then checking throughout the structure for signs of a pungent odor or smoke. This test should be performed by a plumber.

- **House trap.** Some communities require the installation of a house trap on the building drain. This trap is usually located inside the building by the foundation wall. It is U-shaped and requires a separate vent that terminates outside the foundation wall. Inspect the trap cleanout and check to see that the vent is unobstructed from the outside.

Figure 6.3
Assessing Hot Water Heater Capacity

Water heater capacity is determined by the heater's storage capacity and its recovery rate, or the time it takes to reheat the water in its tank. Recovery rates vary with the type of fuel used. Generally, gas- or oil-fired heaters have a high recovery rate and electric heaters have a low recovery rate. Low recovery rates can be compensated for by the provision of larger storage capacity.

Water heaters are sized according to the number of people living in the house and the type of heat source used:

Gas

30 gallon (115 L)	3 to 4 people
40 gallon (150 L)	4 to 5 people
50 gallon (190 L)	5 and more people

Electric

40 to 42 gallon (115 to 160 L)	3 to 4 people
50 to 52 gallon (190 to 200 L)	4 to 5 people
65 gallon (250 L)	5 and more people

Oil

30 gallon (115 L)	any number of people

A qualified plumber or mechanical engineer should determine the size of replacement units based on rehabilitation plans.

If a spa or whirlpool bath is in the house and the water is heated by either gas or electricity, an additional capacity of 10 gallons (40 L) is needed.

6.4 Tank Water Heaters

Tank water heaters consist of a glass-lined or vitreous enamel-coated steel tank covered by an insulated sheet metal jacket. They are gas-fired, oil-fired, or electrically heated.

- **Gas-fired tank water heaters** have an average life expectancy of about 11 to 13 years and a high recovery rate.
- **Oil-fired heaters** have an average life similar to that of gas-fired heaters. Their recovery rate is also high.
- **Electric water heaters** have a longer service life—about 14 years. They have a low recovery rate and thus require a larger storage tank.

Dates of tank manufacture are usually listed on the data plate (often in a simple 1995 code in the serial number; 0595, for instance, would mean manufactured in May 1995), and since water heaters are usually installed within several months of manufacture, the age of the tank often can be approximated. Plan to replace a tank near the end of its life expectancy. Assess water heater capacity in accordance with Figure 6.3. Inspect tank water heaters as follows:

- **Plumbing components.** Check that the hot and cold water lines are connected to the proper fittings on the tank; often they are reversed, causing a loss of fuel efficiency. There should be a shutoff valve on the cold water supply line. Heavy mineral or rust

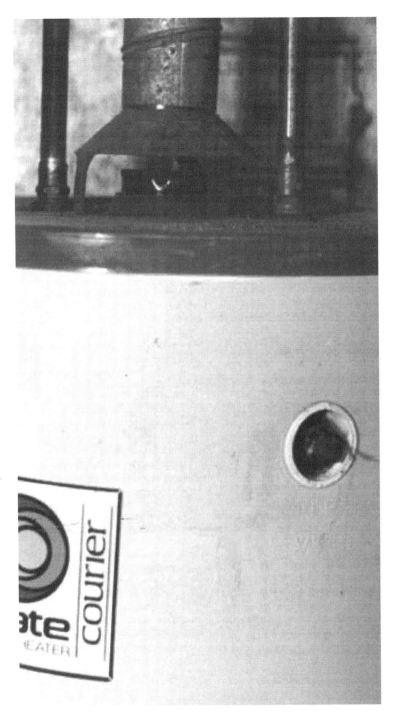

This water heater tank had a threaded plug where its temperature-pressure relief valve should have been, an unsafe condition that should be corrected immediately.

deposits around the tank fittings are usually a sign that the tank is nearing the end of its service life.

Test: If the tank's age or general condition cannot be determined from observation, consider having a plumber drain some water from the tank and inspect for sediment and rust.

Check for signs of leakage on the bottom of the tank, such as rust or water stains on or near fuel burning components or on the floor. Leaking tanks cannot usually be repaired and, therefore, must be replaced entirely. Heavy rusting of the tank interior indicates that the tank should be replaced, although the presence of some sediment and rust is normal. The tank should be drained regularly to remove this normal amount of sediment and rust. Check for the existence of a temperature/pressure relief valve on top of the tank or on the hot water line leading from the tank (it should not be on the cold water line), and for a discharge pipe that extends from the valve to a few inches from the floor or to a floor drain or the building exterior, depending on local code requirements.

Test: If necessary, the pressure relief valve can be tested by pressing the test lever, but since it may stick open, do not perform this test without having a replacement valve available and the necessary tools for replacement.

The soot that has accumulated below the draft hood of this water heater indicates a severely clogged flue or chimney, or more commonly, back-drafting caused by insufficient make-up air.

- **Fuel-burning components.** On gas- and oil-fired tank water heaters, check the flue for upward pitch (1/4 inch per foot [1:50] minimum with no flat spots), tightness of the fittings, and overall condition and integrity. A clogged flue or chimney will deposit soot on top of the tank under the draft hood. On oil-fired water heaters, check for a barometric damper in the flue and verify that it moves freely.

  Test: Check that the burners have an adequate supply of combustion air by using a draft gauge or match to test the draft.

  Inspect the ignition components and look for clogged burners and signs of flashback, such as soot on the heater near the burner. The fuel burning components can be repaired or replaced, but consider the cost-effectiveness of such repairs in terms of the expected remaining service life of the water heater.

- **Seismic vulnerability.** If the building is in seismic zones 3 or 4 (California and portions of Alaska, Arkansas, Hawaii, Idaho, Missouri, Montana, Nevada, Oregon, Utah, Wyoming, and Washington), check the water heater for the presence of seismic bracing to the floor or other structural member.

6.5 Tankless Coil Water Heaters (Instantaneous Water Heaters)

Tankless coil water heaters consist of small diameter pipes coiled inside of or in a separate casing adjacent to a hot water or steam boiler. They are designed for a specific rate of water flow, usually three to four gallons per minute. Since demand for domestic hot water can easily exceed this flow, such heaters often have an associated storage tank to satisfy periods of high demand. Thus the recovery rate of a tankless coil water heater is instantaneous for low demand and will vary for high demand depending on the size of the storage tank, if any. The life expectancy of a tankless coil water heater is limited only by the possible long-term deterioration of its coils and by the service life of the boiler to which it is attached. Since the boiler must operate through the summer in order for the water heater to function, such water heaters are usually considered inefficient.

Check tankless coil water heaters as follows:

- **Plumbing components.** Inspect the plumbing connections and joints around the heater mounting plate for rust, water stains, and mineral deposits. Tighten the mounting plate and repair the connections if required.

  Test. Turn on the hot water in two or more plumbing fixtures to check the water flow. If it is low, suspect a buildup of mineral deposits within the coil. Such deposits can often be flushed from the coil by a plumber.

- **Controls.** Inspect the functioning of the aquastat (device that activates the boiler when heat is needed for producing hot water).

  Test: Run hot water until the boiler fires. Boiler water temperature should not drop below 180 °F (82 °C) on the water gauge; if it does, the aquastat needs adjustment.

  Check for the presence of a pressure relief valve on the hot water side of the coil or on the auxiliary storage tank. The valve should be connected to a discharge pipe that extends to a few inches from the floor, or the building's exterior, depending on local code requirements.

  Test: The relief valve should be tested by pressing the test lever, but as it may stick in the open position, the test should not be performed without having a replacement valve available and the necessary tools for replacement.

6.6 Water Wells and Equipment

Assess well capacity as described in Figure 6.4. Check water wells and equipment as follows:

- **Location and water quality.** Wells that supply drinking water should be located uphill from the building supplied and from any storm or sanitary sewer system piping. Codes usually require that the well be a minimum of 50 feet (15 m) from a septic tank and 100 feet (30 m) from any part of the absorption field; however, local codes may have different separation distances based on the percolation rates of the local soils. Well water can be more corrosive than city water and may contain radon.

Test: Water should be analyzed for the presence of bacterial contamination, for its mineral content, and for the presence of radon. The local health department normally will provide such an analysis. There should be no measurable coliforms.

- **Depth and casing.** Most localities now require wells to be more than 50 feet (15 m) in depth and encased in a steel, wrought iron, or plastic pipe. The casing should extend several inches above its surrounding concrete cover, which should slope away from and completely protect the casing. The casing should be tightly sealed where the pump and power lines enter it and protected from flooding and other threats to its sanitary integrity.

- **Pumps.** Two kinds of deep well pumps are in common use, the jet pump and the submersible pump. A jet pump is mounted above the well casing, and two pipes should extend into it; if there is only one pipe leading into the casing, the well is less than 25 feet deep and may not meet code. Submersible pumps are located at the bottom of the well casing (submerged) and a single discharge pipe and an electrical supply cable extend from the top of the casing. The life expectancy of deep well pumps is 10 years or longer, depending on the type. Submersible pumps are usually the most long lasting and trouble free. Check that pump and plumbing components at the well are protected from freezing.

- **Pressure tank and switch.** A tank under low air pressure (a hydropneumatic tank) should be located in either the well house or the building's basement. This tank regulates water pressure and flow; when air pressure is lost (as air is absorbed in the water over time), the tank becomes waterlogged and causes the pump to be activated every time water is used. Look for this condition; it can be remedied by pumping air back into the tank. Newer tanks contain an air bag. A pressure switch on the tank keeps the water pressure within a predetermined 20 psi (140 kPa) range (usually 20 to 40 psi [140 to 275 kPa], 30 to 50 psi [205 to 345 kPa], or 40 to 60 psi [275 to 415 kPa]).

Test: Check the pressure tank and switch by running the water and seeing whether the pump activates at the lower pressure limit and stops at the upper pressure limit. If pressure slowly goes down in the tank without water being drawn from the system, the tank or some other part of the system is leaking, and the problem should be found and corrected.

Pressure tanks and switches have an average life expectancy of 5 to 10 years, but may last much longer. Check the tank for the presence of a pressure relief valve.

Figure 6.4
Assessing Well Capacity

A water well serving a single-family residence should be capable of sustaining at least a 4-gallon-per-minute flow (a 5- to 7-gpm [19 to 26 L/min] flow is preferable) with a peak flow capacity of 12 gpm (45 L/min).

Test: To check the well's capacity, run water simultaneously from several faucets for 30 minutes or more. Note pressure fluctuations, if any. Near the end of the test, look for mud or cloudiness in the water; this indicates that the well has insufficient capacity for normal use.

Wells serving more than one residence should have proportionately larger capacities. A more exacting capacity test can be performed by a well specialist.

6.7
Septic Systems

Assess septic system capacity as described in Figure 6.5. Check septic systems as follows:

- **Location and layout.** Septic systems should be located downhill from the building. No storm water should be directed into the septic system, as this can flood it and force solids into the absorption field, thereby destroying the field. Sufficient room should exist on the property to relocate the absorption field, which has an average life expectancy of 20 to 30 years under proper use. Do everything possible to determine the layout of the existing septic system, as the absorption field should not be disturbed by new construction and vehicular traffic, or covered by fill. The field often can be located by the presence of greener vegetation in dry summer weather or by melting snow in winter.

- **Septic tank.** The septic tank should be watertight and, for a single-family house, have a minimum capacity of 1000 gallons (3800 L). If properly maintained, it should have been pumped every several years. Ask to see the tank's pumping records. Lack of periodic pumping will cause solids to be carried into the absorption field, clogging the leaching beds and shortening their useful life.

- **Absorption field.** The absorption field should be adequately sized to handle its service loads without clogging or overflowing. Try to locate the original design information and service records for the system. If the company that serviced the installation can be identified,

---

Figure 6.5
Assessing Septic Capacity

Plumbing codes normally require the following septic tank capacities (check the local code for exact requirements):

Single-family, number of bedrooms	Multi-family, one bedroom each	Capacity
1 to 2	--	750 gallons (2840 L)
3	--	1000 gallons (3785 L)
4	2 units	1200 gallons (4545 L)
5 to 6	3 units	1500 gallons (5680 L)
	4 units	2000 gallons (7570 L)

The capacity of the septic system's absorption field depends on its layout and on the percolation qualities of the surrounding soil.

> Test: The absorption field's capacity should be checked by running water into several plumbing fixtures for 30 to 60 minutes and observing the trap in the lowest building fixture. If the water in the trap "boils," backs up, or makes a gurgling sound, the absorption field is clogged or inadequately sized. In either case, it will probably have to be replaced. See a local septic system specialist to determine replacement needs. It is very difficult to test septic systems where the house has been vacant for 30 days or more.

Other signs of a clogged absorption field are the presence of dark green vegetation over the leaching beds throughout the growing season (caused by nutrient-laden wastes being pushed up through the soil), wet or soggy areas in the field, or distinct sewage odors. These signs all indicate the probable need to replace the absorption field.

Residential Inspection 81

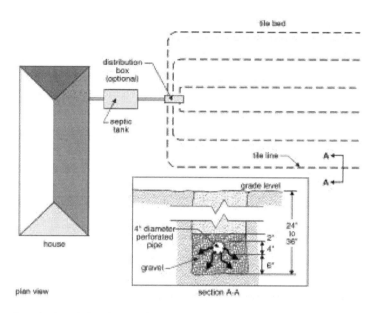

Example of a typical septic system layout with a detail of the absorption field piping

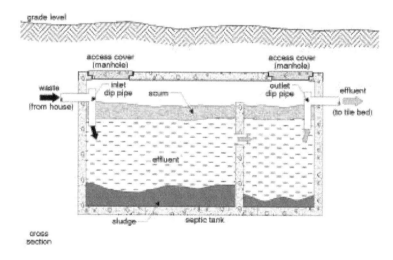

Cross section of a typical two-compartment septic tank

additional work is needed. A qualified plumber or mechanical engineer should determine the capacity of the system based on the renovation plans.

- **Grease trap.** Some houses have grease traps in the septic system to prevent grease from getting into and clogging the absorption field. This trap should be inspected for grease buildup.

Test: To check whether runoff on the site is coming from the absorption field, put dye capsules in the waste water and return later to check the color of any runoff.

6.8 Gas Supply in Seismic Regions

Inspect the following features of the gas service:

- **Service entrance.** If the building is in seismic zones 3 or 4 (California and portions of Alaska, Arkansas, Hawaii, Idaho, Missouri, Montana, Nevada, Oregon, Utah, Wyoming, and Washington), check the gas service for vulnerability to differential movement where the piping enters the building. Look for adequate clearance or for flexible connections.

- **Emergency shutoff.** If the building is in seismic zone 4 (portions of Alaska and California, and small parts of Idaho, Montana, and Wyoming), look for an automatic emergency shutoff valve for the entire house.

check with that company about the system's condition. If a renovation that adds bedrooms is being considered, verifying the system's existing capacity is critical to determining what

7
HVAC System

Most HVAC (heating, ventilating, air conditioning) systems in small residential buildings are relatively simple in design and operation. They consist of four components: controls, fuel supply, heating or cooling unit, and distribution system. Each component must be evaluated for its physical and functional condition and its adequacy in terms of the building's planned reuse. The adequacy of heating and cooling is often quite subjective and depends upon occupant perceptions that are affected by the distribution of air, the location of return air vents, air velocity, the sound of the system in operation, and similar characteristics. For this reason, past energy use should not be used as the basis for estimating future energy use.

This chapter describes inspection procedures for oil- and gas-fired warm air, hot water, and steam heating systems; electric resistance heaters; chilled air and evaporative systems; humidifiers; unit air conditioners; and attic fans.

When inspecting the HVAC system, look for equipment service records and read all equipment data plates. Whenever possible, ask building occupants about the HVAC system's history of performance. Always try to observe equipment in actual operation.

When universal design is a part of a rehabilitation, consult HUD publication *Residential Remodeling and Universal Design* for detailed information about HVAC controls.

HVAC systems have used asbestos-bearing insulation on piping, ducts, and equipment, and may have lead-based paint on piping and equipment such as radiators. When inspecting the HVAC system, pay particular attention to the presence of these hazardous materials.

Assess heating and cooling capacity as described in Figure 7.1.

7.1
Thermostatic Controls

Residential HVAC controls consist of one or more thermostats and a master switch for the heating or cooling unit. Inspect them as follows:

■ **Thermostats.** Thermostats are temperature-sensitive switches that automatically control the heating or cooling system. They normally operate at 24 volts. Thermostats should be located in areas with average temperature conditions and away from heat sources such as windows, water pipes, or ducts. For a thermostat that controls both heating and cooling, a location near the return air grille is ideal.

Test: Check each thermostat by adjusting it to activate the HVAC equipment. Then match the temperature setting at which activation occurs with the room temperature as shown on the thermostat's thermometer.

Take off the thermostat cover and check for dust on the spring coil and dirty or corroded electrical contact points.

Newer thermostats have a mercury switch in lieu of electrical contacts. Plan to replace worn or defective thermostats.

There may be more than one thermostat in each living unit. Sometimes two thermostats separately control the heating and cooling system, and sometimes the living unit is divided into zones, each with its own thermostat. Multi-family buildings with a central HVAC system will be divided into at least one zone per living unit and buildings with electric baseboard heat may have a thermostat in every room or on every heating unit.

Test: Check the functioning of multi-zone systems by operating the HVAC system in all its modes and noting whether distribution is adequate in each zone (see also Sections 7.3 and 7.4). Consider the zoning needs for the planned rehabilitation of the building. Refer to the National Environmental Balancing Bureau's *Procedural Standards for Testing, Adjusting, and Balancing of Environmental Systems* or the Associated Air Balance Council's *MN-4, ABBC Test and Balance Procedures*.

■ **Master switch.** Every gas- and oil-burning system should have a master switch that serves as an emergency shutoff for the burner. Master switches are usually located near the burner unit or, if there is a basement, near the top of the stairs.

Cooling system controls also may include a master switch, which in the "off" position will not allow the compressor to start, as well as a switch allowing only the circulating fan to operate.

> ### Figure 7.1
> ### Assessing Heating and Cooling Capacity
>
> The capacity of an existing heating or cooling system, as measured by its ability to heat or cool a specific building or space, can be determined in either of two ways:
>
> - **Field test.** Properly sized heating and cooling systems should operate at full capacity at normal yearly outside temperature extremes and should be slightly undersized for unusual outside temperature extremes. It is rare, however, that they can be checked under such conditions.
>
>   Test: Operate the heating system on the coolest possible day and the cooling system on the warmest possible day (within the limitations of the inspection period). Note how "hard" the system is working to maintain the preset indoor temperature, as indicated by how often the system cycles on and off, and compare this to outside temperatures. This procedure, while inexact, may provide some idea of the system's potential capacity.
>
>   When the system has a history of continuous use, maintenance, and repair, it can be assumed to have sufficient capacity. However, check with present or former building tenants on this matter.
>
>   Of more concern is the fuel efficiency of the system. Ask the local utility company or fuel distributor for records of past fuel consumption and consider this in the overall assessment of the HVAC system.
>
> - **Design calculation.** An HVAC system's capacity can be more accurately determined by noting its heating or cooling output (in tons or BTUs) from information on the manufacturer's data plate and comparing it to the building's heating and cooling loads. These loads can be calculated using the Air Conditioning Contractors of America's Manual J or similar load calculation guide.
>
>   A rough estimate of a building's required heating equipment size in BTUs per hour (BTUH) can be obtained by using the following formula:
>
>   BTUH = .33 x [square footage of building to be heated] x [difference between outside and inside design temperatures]
>
>   The factor of .33 in this formula is based on R11 exterior walls, an R19 ceiling at the top floor or roof, and double-glazed windows.
>
>   A rough estimate of a building's required cooling equipment size, in tons, can be made by dividing the floor area by 550 (each ton equals 12,000 BTUH). Tonnage is not an adequate measure of cooling capacity in a dwelling of three or more floors with the air handling unit located on the lowest floor, with such a layout, the top floor can never be properly cooled.
>
>   These estimates should be followed by a complete load calculation after rehabilitation needs are firmly established.

Test: Operate all master and emergency shut-off switches when the burner is in operation to see whether they deactivate the unit.

In hot water heating systems that also are used to generate domestic hot water, the thermostat controls the circulating pump rather than the burner (see Section 7.4).

7.2 Fuel-Burning Units, General

Oil- or gas-fired furnaces and boilers provide heat to the majority of small residential buildings. Such fuel-burning units, whether they are part of a warm air or a hot water system, should be inspected as follows:

- **Location, clearances, and fire protection.** Check that the unit meets local fire safety regulations. No fuel-burning unit should be located directly off sleeping areas or close to combustible materials.

- **Data plate and service records.** Locate the data plate on each unit and note its date of manufacture, rated heating capacity in BTUs per hour, fuel requirements, and other operational and safety information. Examine the service records of oil-fired units. These should be attached to the unit or available from the oil distributor or company that last serviced the unit.

- **Seismic vulnerability.** If the building is in seismic zones 3 or 4 (California and portions of Alaska, Arkansas, Hawaii, Idaho, Missouri, Montana, Nevada, Oregon, Utah, Wyoming, and Washington), check fuel-burning equipment for the presence of seismic bracing to the structure.

- **Fuel supply.** Gas supply lines should be made of black iron or steel pipe (some jurisdictions allow copper lines with brazed connections). Shutoff valves should be easily accessible and all piping well-supported and protected.

Oil tanks should be maintained in accordance with local code or the recommendations of the National Fire Protection Association. All tanks must be vented to the outside and have an outside fill pipe. Buried

Record all pertinent information from the manufacturer's data plates on HVAC equipment. It will be useful in assessing the equipment's capacity later.

tanks normally have a 550, 1000, or 1500 gallon (2080, 3785, or 5680 L) capacity; basement tanks are usually restricted to a 275 gallon (1040 L) capacity, with no more than two tanks allowed. Tanks must be located at a minimum of seven feet from the furnace and should be adequately supported and free of interior rust. Outside tanks at grade should have an adequate supporting base.

Oil tanks often begin to leak after about 20 years, when the bottom of the tank corrodes from moisture that has condensed inside the tank and settled to the bottom. Feel along the undersides and probe the interiors for such leakage. Look for an oil level gauge and see whether it works. Decide whether the tanks should be replaced. See also the information on buried oil tanks in Section 1.2.

Check the oil supply line to the furnace; it should be equipped with a filter and protected from accidental damage and rupture.

■ **Ventilation and access.** Make sure the fuel-burning unit has adequate combustion air and is easily accessible for servicing with at least three feet clear on each side of the unit requiring service. Check the local code for requirements. Also check equipment manufacturer's guidelines for makeup air, especially where furnaces and boilers are enclosed in a finished basement or closet. A general rule is to provide one inch of free area across the width of the door to the furnace or boiler room or closet for every 1000 BTUH (300 W) of heating. The free area needed should be divided: roughly half at the bottom of the door and half at the top. A grille can also be used in the door.

■ **Condition.** Open all access panels and examine the external and internal condition of each unit. On hot air furnaces, look for signs of rust from basement dampness or flooding, and, if an air conditioning evaporator coil is located over the furnace, look for rust caused by condensate overflow. On hot water boilers, look for rust caused by dampness and by leaking water lines and fittings. If possible, check the condition of the interior refractory lining on all oil-fired units.

■ **Ignition and combustion.** Observe the ignition and combustion process.

Test: Step away from the unit while someone else turns up the thermostat. Look for a puffback in oil-fired units or flames licking under the cover plate of a gas-fired unit; both indicate potential hazards that must be corrected. If the unit doesn't light, check the master switch or emergency

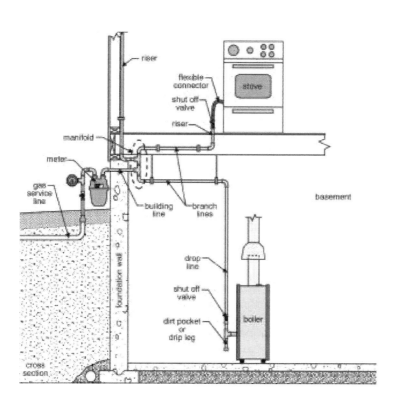

Gas piping terminology

Look for signs of corrosion around and within oil storage tanks and check the operation of the oil level gauge. Use a dipstick to check for signs of condensation in the tank.

shutoff to make sure it's on, press the reset button, and try again. If it still doesn't light, call a service technician.

Once the unit has been activated, closely observe the combustion process. In oil-fired units, the flame should be clear and clean, and have minimal orange-yellow color. Flame height should be uniform.

Gas-fired units should have a flame that is primarily bluish in color. Note whether the flame lifts off the burner head; this indicates that too much air is being introduced into the mixture. Check gas burners for rust and clogged ports. Soot build up is a sign of inefficient combustion. In oil-fired units, look for soot below the draft regulator, on top of the unit's housing, and around the burner. The odor of smoke near the unit is another sign of poor combustion.

Test: Consider having a service technician perform a flue gas analysis to determine the unit's combustion efficiency. This test requires the use of a flue gas analyzer and should be performed in accordance with ASTM D2157, *Standard Test Method for Effect of Air Supply on Smoke Density in Flue Gasses from Burning Distillate Fuels.*

- **Venting and draft.** Check the smoke pipe between the unit and the chimney. It should have a slight upward pitch with no sags, preferably a minimum of 1/4 inch per foot. Inspect the pipe for corrosion holes, the tightness of its fittings, and the tightness of its connection to the chimney. Check for signs of soot build up in the smoke pipe. Consult local code requirements about the minimum size, required clearance from combustible materials, and number of

This barometric draft regulator should swing freely and open somewhat as the heating unit warms up.

smoke pipes entering the chimney. Newer, higher efficiency furnaces are not as prone to backdrafting because of forced or reduced draft systems. When these systems are used with existing old flues, flues tend to fail early. Check for evidence of rust or leaking in the exhaust flue.

Gas-fired units have a draft diverter that is located either on the exhaust stack of a boiler or built into the sheet metal casing of a furnace.

Test: Have a service technician run the furnace or boiler through a complete cycle, then with a match or candle conduct a simple smoke test of the draft at or near the diverter. A draft gauge or CO tester can be used to detect an outward flow of hot exhaust gas; this indicates a hazardous draft problem that must be corrected.

Proper draft is critical to the efficient operation of an oil-fired unit. A barometric draft regulator is required above the unit or on the smoke pipe. Inspect for open joints or cracks that allow excess air to enter the combustion chamber or the smoke pipe. All such openings should be sealed. The damper of a barometric draft regulator should be level, free of rust, and not damaged or altered. Improper draft from an oil furnace could cause a build up of carbon monoxide gas in occupied spaces. Have old flues cleaned by a chimney sweep or HVAC service technician. Have a deteriorated flue replaced.

Test: Check the draft regulator by observing its motion when the heating unit is in operation. It should open as the heating unit warms up. The draft regulator is adjusted during the combustion efficiency test.

■ **Operation.** The operation of the fuel-burning unit will depend on the type of heating system in which it is used. See Section 7.3 for the operation of gas- and oil-fired warm air systems and Section 7.4 for the operation of gas- and oil-fired hot water and steam systems.

7.3 Forced Warm Air Heating Systems

Warm air heating systems are of two types, forced air or gravity. Gravity systems are occasionally still found in older single-family houses, but most gravity systems either have been replaced or converted to forced air. Gravity systems are big, bulky, and easily recognizable. Lacking a mechanical means of moving air, such systems are inefficient and heat unevenly, can be dangerously hot, and are generally considered archaic. Plan to replace them unless there are overriding reasons for doing otherwise.

Most forced warm air systems use natural gas or fuel oil as a heat source, but some systems use electric resistance heaters or heat pumps. These heaters replace the heat exchanger and burner found in gas- and oil-fired furnaces or supplement the heat output of heat pumps (see Section 7.9). Electric resistance heating systems have no moving parts and require no adjustment. The circulation blower and air distribution ductwork for electric resistance heating systems (and heat pumps) are identical to those of gas- and oil-fired warm air systems and should be checked as described below. See Section 7.6 for additional information on electrical resistance heating equipment.

Assess the condition of forced warm air heating systems as follows:

■ **Heat exchanger.** The heat exchanger is located above the burner in gas- and oil-fired furnaces and separates the products of combustion from the air to be heated. (There is no heat exchanger in an electrically heated furnace.) It is critical that the heat exchanger be intact and contain no cracks or other openings that could allow combustion products into the warm air distribution system. Visual detection of cracks, even by heating experts, is a difficult and unreliable process.

Test: Look for signs of soot at supply registers and smell for oil or gas fumes. Observe the burner flame as the furnace fan turns on; a disturbance or color change in the flame may indicate air leakage through the exchanger. Operate the furnace for several minutes and then feel the furnace frame for uneven hot spots. Similarly, another simple test requires turning on the fan only and placing a lighted match or candle in the heat exchanger enclosure. If there are leaks, the flame will flicker. A CO tester may also be used to detect combustion gases. For any of these tests, consult a heating contractor or HVAC service technician.

Look for rust on the exchanger— a major cause of premature exchanger failure is water leakage from humidifiers or blocked air conditioner condensate lines. Check for other signs of water leakage.

The durability of the heat exchanger determines the service life of the furnace. Furnaces installed since the 1950s normally have a useful life of 25 years or less. Older furnaces with cast iron heat exchangers may last much longer.

■ **Furnace controls.** Gas- and oil-fired furnaces have two internal controls, a fan control and a high-temperature limit control. (Furnaces with electric resistance heating coils have high temperature limit controls and air flow switches.) The fan control prevents cold air from being circulated through the system. It is a temperature-sensitive switch, completely independent of the thermostat, and turns the furnace blower on and off at preset temperatures. When the thermostat calls for heat, the furnace burner is turned on. After the heat exchanger warms to a preset temperature (usually 110 to 120 °F [43 to 49 °C]), the fan control activates the blower. The thermostat will shut off the burner when the building warms to the thermostat setting, and when the heat exchanger cools to about 85 °F (29 °C) the fan control will switch off the blower.

Test: Observe the above sequence; if it is faulty, the fan control should be adjusted or replaced.

The high-temperature limit control is a safety device that shuts the burner off if the heat exchanger gets too hot (the control is usually set at about 175 °F). Should the burner automatically turn off before the blower is activated, either the blower, the fan control, or the high-temperature limit control is faulty and should be adjusted or replaced.

■ **Circulation blower.** Remove the blower cover and inspect the blower motor and fan. Look for proper maintenance and oiling. Check for wear or misalignment of the fan belt, if any, and for dirt build up on the motor or fan.

Test: When the system is operating, listen for unwarranted blower noise and determine its cause.

■ **Distribution system and controls.** The distribution system is made up of supply and return ducts, filters, dampers, and registers. Supply and return ducts may be made of sheet metal, glass fiber, or other materials. Glass fiber ducts are self-insulated, but sheet metal ducts are usually not insulated except where they pass through unheated (or uncooled) spaces (see Sections 3.1 and 3.9). Sheet metal ducts are occasionally insulated on the inside; determine the presence of insulation by tapping on the duct and listening for a dull sound. Check ducts for open joints and air leakage wherever the ducts are exposed. Examine them for dirt build up by removing several room registers and inspecting the duct. Ducts can be cleaned by a heating contractor. If there is a flexible connection between the furnace and the duct work, check it for tears and openings. There should be no openings in return ducts in the same room as a combustion furnace.

Air filters are usually located on the return side of the furnace next to the blower, but they may be found anywhere in the distribution system. Check for their presence and examine their condition.

Supply ducts are often provided with manual dampers to balance air flow in the distribution system. Locate them by looking for small damper handles extending below the ductwork. Check their operation. In zoned systems, automatically controlled dampers may be located in the ductwork, usually near the furnace.

Test: The operation of all dampers should be checked by activating each thermostat, one at a time. If the dampers are working properly, air should begin to circulate in each zone immediately after its thermostat has been activated.

Check the location of supply and return registers in each room. Warm air registers are most effective when positioned low on the exterior wall; cold air registers when located high on the walls or in the ceiling. Return registers should be on opposite sides of the room from supply registers. If return registers are located in a hallway or a different room, make sure intervening doors are undercut by about one inch.

Test: When the furnace blower is on, check the air flow in all supply and return registers. Remove and inspect registers that appear blocked. Listen for sounds emanating from the ductwork and determine their source.

Humidifiers may be located in the supply ducts. They should not be located in return air ducts because the moist air will pass through the heat exchanger and evaporator coil, rendering the humidification ineffective and corroding the heat exchanger. Check humidifiers in accordance with Section 7.11.

7.4 Forced Hot Water (Hydronic) Heating Systems

Hot water heating systems, like warm air systems, are of two types, forced or "hydronic" and gravity. Gravity systems are sometimes found in older single-family houses, but in most cases such systems have been replaced or converted to a forced hot water system. Gravity systems have no water pump and use larger piping. They tend to heat unevenly, are slow to respond, and can only heat spaces above the level of their boiler. Like gravity warm air systems, they are considered inefficient and normally should be replaced during the rehabilitation process.

Forced hot water systems are usually heated by gas- or oil-fired boilers. Occasionally they may use immersion-type electric resistance heating coils. These coils replace the burner found in gas- and oil-fired boilers. The hot water pump and distribution piping for electrically heated systems are similar to those of gas- and oil-fired hot water systems and should be checked as described below. Refer to Section 7.6 for additional information on electrical resistance heating equipment.

Assess the condition of forced hot water heating systems as follows:

- **Boiler.** Most hot water and steam heating systems have steel boilers with a service life of about 20 years. Cast iron boilers, which are less

 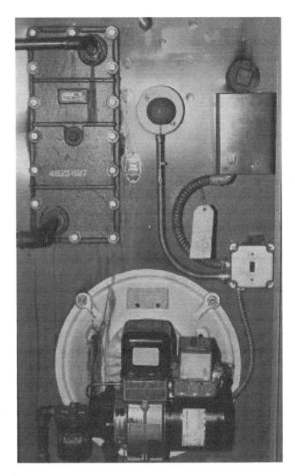

The hot water boiler on the left is gas-fired; the one on the right, oil-fired. Both have standard pressure and temperature gauges.

common, have a service life of about 30 years. Old cast iron boilers converted from coal-fired units may last much longer but are usually quite inefficient. Inspect all boilers for signs of corrosion and leakage.

Test: Run the boiler for one-half hour or longer and check for leaks. Occasionally a boiler fitting will leak slightly before it warms up, expands, and returns to a watertight fit. Don't confuse condensation droplets on a cold boiler with water leaks.

- **Expansion tank.** The expansion tank is usually located above the boiler (although it may be in the attic) and is connected to the hot water distribution piping. Most tanks are compression-type tanks that are designed to permit heated water to expand against a cushion of pressurized air within the tank. When the tank loses air, it becomes "waterlogged" and expansion cannot be accommodated. Instead, water discharges from the boiler's pressure relief valve each time the system heats up. Check for such a condition. Waterlogged expansion tanks should be drained and repressurized. This should be done by a heating or plumbing contractor.

- **Boiler controls.** All boilers should be equipped with a pressure gauge, a pressure relief valve, and a pressure-reducing valve. The pressure gauge indicates the water pressure within the boiler, which should normally be between 12 and 22 psi (83 to 153 kPa). A temperature gauge may be included in the pressure gauge. The pressure-reducing valve (actually a water make-up valve) adds water to the system from the domestic water supply when the boiler pressure drops below 12 psi (83 kPa). Pressure readings lower than 12 psi indicate a faulty valve that should be adjusted or replaced.

The pressure relief valve should discharge water from the system when the boiler pressure reaches 30 psi (207 kPa). Look for signs of water near the valve or below it on the floor. High pressure conditions are usually due to a waterlogged expansion tank. If the boiler also generates domestic hot water, high pressure may be caused by cracks in the coils of the water heater, since the domestic water supply pressure usually exceeds 30 psi (207 kpa). The pressure relief valve should be mounted on the boiler.

Test: As a last resort, the pressure relief valve may be tested by a service technician. But since it may be old or clogged and become stuck in the open position, the test should not be performed without having a replacement valve on hand and the proper tools for removing and reinstalling the valve and extension pipe.

Hot water boilers should have a high-temperature limit control or aquastat that shuts off the burner if the boiler gets too hot. Check for such a control.

■ **Circulating pump and controls.** The circulating pump forces hot water through the system at a constant flow rate, usually stated in gallons per minute (gpm). It should be located adjacent to the boiler on the return pipe near the boiler. The pump may be

This hot water expansion tank is located above the boiler. Look for signs of leakage in expansion tanks.

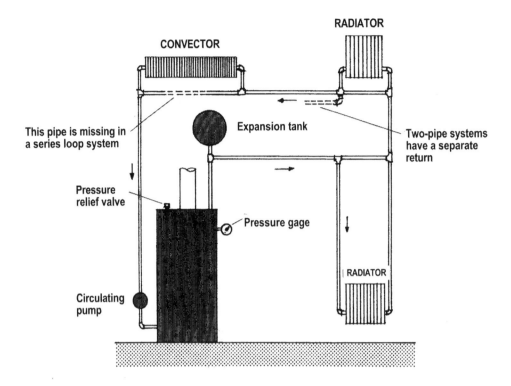

One-pipe forced hot water heating system

operated in one of the following four ways:

❑ *Constant-running circulator,* in which the pump is controlled by a manual switch. The pump is usually turned on at the beginning of the heating season and runs constantly until it is turned off at the end of the heating season. The boiler is independently activated by the thermostat as heat is required.

❑ *Aquastat-controlled circulator,* which turns the pump on and off at a preset boiler temperature (normally 120 °F [49 °C]). Like the constantly running circulator, the burner is independently activated by the thermostat as heat is required.

❑ *Thermostat-controlled circulator,* in which water is maintained at a constant temperature in the boiler by an aquastat.

❑ *Relay-controlled circulator,* in which the pump is activated (via a relay switch) whenever the boiler is activated by the thermostat.

Test: Determine which kind of device controls the pump and check its operation. Inspect the condition and operation of the pump itself. Listen for smooth operation; a loud pump may have bad bearings or a faulty motor. Inspect the seal between the motor and the pump housing for signs of leakage. Examine the condition of all electrical wiring and connections. Feel the return line after the system has been operating for a short time; it should be warm. If it isn't, the pump may be faulty.

In heating systems used for generating domestic hot water, the thermostat will control the circulating pump and an aquastat will control the burner. See Section 6.5.

■ **Distribution piping.** The forced hot water distribution system consists of distribution piping, radiators, and control valves. Distribution piping may

The hot water radiator on the left has a bleed valve at its top left corner. The steam radiator on the right has an air vent.

be one of three types: series-loop, one-pipe, and two-pipe.

In a **series-loop system**, radiators are connected by one pipe directly in a series. Since the last radiator will receive cooler water than the first, downstream radiators should be progressively larger. Alternatively, series-loop systems may be divided in small zones to overcome this problem.

One-pipe systems differ from series-loop systems in that their radiators are not connected in series. Instead, each radiator is separately attached to the water distribution pipe with a diverter fitting, which is used to regulate the amount of hot water entering it.

Two-pipe systems use separate pipes for supply and return water, which ensures a small temperature differential between radiators, regardless of their location. Individual room control is possible with both one-pipe and two-pipe systems, although a change in the valve adjustment on one radiator will affect the performance of others downstream.

Distribution piping should be checked for leaks at valves and connections. Inspect such piping as outlined in Section 6.2 for domestic water supply piping. Make sure pipes are properly insulated in unheated basements, attics, and crawl spaces. See Sections 3.1 and 3.9.

When the distribution piping is divided in zones, each zone will have either a separate circulation pump or a separate electrically operated valve.

Test: Check the operation of all zone valves by activating each thermostat, one at a time. If hot water is being distributed properly to each zone, the radiators in that zone should be warm to the touch within several minutes. Locate all valves, inspect their electrical wiring and connections, and look for signs of leakage.

- **Radiators and control valves.** Radiators are of three types: cast iron (which in most cases are free standing but sometimes are hung from the ceiling or wall, convector (which may have a circulating fan), and baseboard. Older residential buildings usually have cast iron radiators that are extremely durable and normally can be reused, although they are less efficient than convectors. Baseboard radiators are considered the most desirable for residential use because they are the least conspicuous and distribute heat most evenly throughout the room. Radiators should be located on outside walls whenever possible.

Test: Activate the system and look for signs of water leakage. Feel the surfaces of all radiators to ensure that they are heating uniformly; if they are not, bleed them to remove entrapped air. Examine the fins of all convectors for dirt and damage. These fins can be "combed" straight. Check the condition of all radiator control, safety, and bleed valves and make sure they are operational. Often the valves need tightening or their packing needs replacing.

- **Radiant panel heating.** Hot water distribution piping may be embedded in floors, walls, and ceilings to provide radiant heating. Because the piping is embedded, it can only be inspected by looking for signs of water leakage or rust on the floor or wall surfaces that cover the piping. Such heating is normally trouble free, unless there are major structural problems that damage distribution piping and joints, or

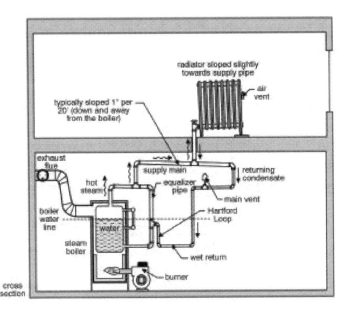

One-pipe parallel flow steam heating system

unless piping includes copper pipe and steel pipe that are not separated by a dielectric coupling.

Test: Operate radiant heating systems to determine their functional adequacy. Radiant surfaces should be warm to the touch within several minutes. Check the condition of the shutoff valves for each distribution zone and the main balancing valves near the boiler and look for signs of leakage. Inspect the expansion tank and air vent (if any) in accordance with the subsection on expansion tanks (above). If the system appears to be in good condition but heating is not adequate, consider having a service technician pressure test the distribution piping for a period of up to 24 hours. If a drop in pressure occurs, there is a leak. When a leak is detected, have the service technician flush the piping to check for galvanic corrosion.

7.5 Steam Heating Systems

Steam heating systems are seldom installed now in small residential buildings but are still common in many older ones. They are simple in design and operation, but require a higher level of maintenance than modern residential heating systems. Unless the steam system is in good working order and adequate plans can be made for its up keep, consider replacing it with a more maintenance-free system.

Assess the condition and operation of steam heating systems as follows:

- **Boiler.** Steam boilers are physically similar to hot water boilers and should be inspected similarly. See Section 7.4.

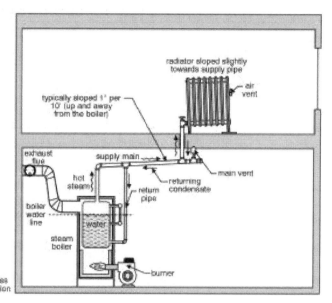

One-pipe counterflow steam heating system

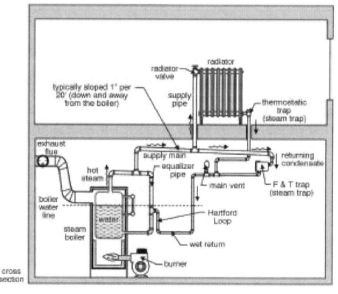

Two-pipe steam heating system

■ **Boiler controls.** Unlike hot water boilers, steam boilers operate only about three-fourths full of water and at much lower pressures, usually 2 to 5 psi. Steam boilers should be equipped with a water level gauge, a pressure gauge, a high-pressure limit switch, a low water cut-off, and a safety valve.

Test: Activate the boiler and observe the water level gauge that indicates the level of the water in the boiler. The gauge should normally read about half full, though the actual level of the water is not critical as long as the level is showing. If the gauge is full of water, the boiler is flooded and water must be drained from the system. If the gauge is empty, the boiler water level is too low and must be filled (either manually through the fill valve or automatically through the automatic water feed valve, if the boiler has one). Unsteady, up and down motion of water in the gauge means the boiler is clogged with sediment or is otherwise operating incorrectly and must be repaired. The clarity of the boiler water should be noted when checking the gauge; if the gauge is too dirty to judge the water level, remove and clean it. This test and any resulting work should be done by a service technician.

The high-pressure limit switch turns off the burner when the boiler pressure exceeds a preset level, usually 5 to 7 psi (35 to 48 kPa). It is connected to the boiler by a pigtail-shaped pipe. The low water cut-off shuts down the burner when the boiler water level is too low.

Test: Lower the water level in the boiler and see whether the low water cut-off turns off the burner. Usually this test should be performed by a service technician.

The pressure relief valve is designed to discharge when the boiler pressure exceeds 15 psi (103 kPa).

- **Distribution piping.** The steam distribution system consists of distribution piping, radiators, and control valves. Distribution piping may have either a one-pipe or two-pipe configuration.

In a **one-pipe system,** steam from the boiler rises under pressure through the pipes to the radiators. There it displaces air by evacuation through the radiator vent valves, condenses on the radiator's inner surface, and gives up heat. Steam condensate flows by gravity back through the same pipes to the boiler for reheating. The pipes, therefore, must be pitched no less than one inch in ten feet in the direction of the boiler to ensure that the condensate does not block the steam in any part of the system. All piping and radiators must be located above the boiler in a one-pipe system.

In a **two-pipe system,** steam flows to the radiators in one pipe and condensate returns in another. A steam trap on the condensate return line releases air displaced by the incoming steam. If the condensate return piping is located below the level of the boiler, it should be brought back up to the level of the boiler and vented to the supply piping in a "Hartford Loop." This prevents a leak in the condensate return from emptying the boiler. Two-pipe systems can be balanced by regulating the supply valves on each radiator, and may be converted for use in a hot water heating system (although new, larger-size return piping usually must be installed).

Distribution piping should be checked for leaks at all valves and connections. Make sure all piping is properly pitched to drain toward the boiler. "Pounding" may occur when oncoming steam meets water trapped in the system by improperly pitched distribution piping or by shutoff valves that are not fully closed or fully open. Inspect the condition of all piping as outlined in Section 6.2 for domestic water supply piping. Make sure pipes are properly insulated in unheated basements, attics, and crawl spaces. See Sections 3.1 and 3.9.

- **Radiators and control valves.** Steam radiators are made of cast iron and are usually free standing. They are quite durable and, in most cases, can be reused. Radiators should be located on outside walls whenever possible.

Test: Activate the system and inspect the condition of all radiators. Look for signs of water leakage. Feel their surfaces to make sure they are heating uniformly; if they are not, check the radiator air vents and supply valves on a one-pipe system and the radiator supply valves and the steam trap on the condensate return on a two-pipe system. Often air vents need cleaning and supply valves need tightening, or valve packing needs to be replaced. "Pounding" near the radiator can often be cured by lifting one edge of the radiator slightly; this reduces condensate blocking in the pipes.

7.6
Electric Resistance Heating

Electric resistance heating elements commonly are used in heat pump systems, wall heaters, radiant wall or ceiling panels, and baseboard heaters. They are less frequently used as a heat source for central warm air or hot water systems. Such heating devices usually require little maintenance, but their operating costs should be carefully considered when planning the building rehabilitation.

Assess the condition of all electric resistance heating devices by activating them and inspecting as follows:

- **Electric resistance heaters.** Electric resistance heaters are used in warm air and hot water systems as described in Sections 7.3 and 7.4, and in heat pumps as described in Section 7.9. They incorporate one or more heavy duty heating elements that are actuated by sequence relays on demand from the thermostat. The relays start each heating element at 30-second intervals, which eliminates surges on the electrical power system. In warm air and heat pump systems, electric heating elements are normally located in the furnace or heat pump enclosure, but they may be located anywhere in the ductwork as primary or secondary heating devices.

Inspect the condition of all electric resistance heaters, including their wiring and connections.

Test: If possible, observe the start up of the heating elements. Their failure to heat up indicates either a burnt-out element or a malfunctioning relay.

- **Electric wall heaters.** These compact devices are often used as supplementary heating units or as sole heat sources in houses for which heating is only occasionally required. They may have one or more electric heating elements, depending on their size, and should be inspected as described above. Wall heaters often have a small circulation fan; check its condition and operation and look for dirt build up on the fan blades and motor housing. Inspect all electrical wiring and connections.

- **Radiant wall and ceiling panels.** Electric heating panels that are embedded in wall or ceiling surfaces cannot be directly inspected, but all radiant surfaces should be examined for signs of surface or structural damage.

Test: If the panels do not provide heat when the thermostat is activated, check the thermostat, circuit breaker, and all accessible wiring to determine the cause or have an electrical continuity test performed by an electrician.

- **Baseboard heaters.** Baseboard heater heating fins can be damaged and become clogged with dust. Remove heater covers and inspect for such problems. Bent heating fins can often be restraightened by "combing." The thermostat may be on an adjacent wall or in the unit itself.

7.7
Central Air Conditioning Systems

Central air conditioning systems are defined here as electrically operated refrigerant-type systems used for cooling and dehumidification. (Evaporative coolers are described in Section 7.10 and gas-absorption systems are described in Section 7.8.) Heat pumps are similar to central air conditioners, but are reversible and can also be used as heating devices; they are described in Section 7.9. Air conditioning systems should be tested only when the outside air temperature is above 65 °F (18 °C); below that temperature, the systems will not operate properly and may shut down due to safety controls.

There are two types of central air conditioning systems: integral and split. In the **integral system**, all mechanized components—compressor, condenser, evaporator, and fans—are contained in a single unit. The unit may be located outside the building with its cold air ductwork extending into the interior, or it may be located somewhere inside the building with its exhaust air ducted to the outside.

In the **split system**, the compressor and condenser are located outside the building and are connected by refrigerant lines to an evaporator inside the building's air distribution ductwork. Split systems in buildings heated by forced warm air usually share the warm air system's circulating fan and ductwork. In such cases, the evaporator is placed either directly above or below the furnace, depending on the furnace design.

Assess the condition of central air conditioning systems as follows:

- **Compressor and condenser.** The compressor pumps refrigerant gas under high pressure through a condenser coil, where it gives up heat and becomes a liquid. The heat is exhausted to the outside air by the condenser fan. Compressors have a service life of 5 to 15 years, depending on the maintenance they receive, and are the most critical component in the air conditioning system.

Test: Activate the system and observe the operation of the compressor. It should start smoothly and run continuously; noisy start up and operation indicates a worn compressor. The condenser fan should start simultaneously with the compressor. After several minutes of operation, the air flowing over the condenser should be warm. If it isn't, either the compressor is faulty or there is not enough refrigerant in the system.

If the compressor, condenser, and condenser fan are part of a split system and are located in a separate unit outside, check the air flow around the outside unit to make sure it is unobstructed. Look for dirt and debris inside the unit,

Looking down into an outdoor compressor unit with the top cover removed. The compressor is on the upper right, controls are on the lower right, and the fan and condenser coils are to the left.

particularly on the condenser coils and fins, and inspect all electrical wiring and connections. The unit should be level and well-supported, and its housing intact and childproof. An electrical disconnect switch for use during maintenance and repairs should be located within sight of the unit.

Integral systems located somewhere on or in the building should have their compressors placed on vibration mountings to minimize sound transmission to inhabited building spaces.

- **Refrigerant lines.** Refrigerant lines form the link between the interior and exterior components of a split system. The larger of the two lines carries low pressure (cold) refrigerant gas from the evaporator to the compressor. It is about the diameter of a broom handle and should be insulated along its entire length. The smaller line is uninsulated and carries high pressure (warm) liquid refrigerant to the evaporator. Check both lines for signs of damage and make sure the insulation is intact on the larger line. On its exterior, the insulation should be protected from ultraviolet damage by a covering or by white paint. Sometimes a sight glass is provided on the smaller line; if so, the flow of refrigerant should look smooth through the glass. Bubbles in the flow indicate a deficiency of refrigerant in the system. Frost on any exposed parts of the larger line also indicates a refrigerant deficiency.

- **Seismic vulnerability.** If the building is in seismic zones 3 or 4 (California and portions of Alaska, Arkansas, Hawaii, Idaho, Missouri, Montana, Nevada, Oregon, Utah, Wyoming, and Washington), check roof-mounted compressor and condenser units for the presence of seismic bracing to the structure.

- **Evaporator.** The evaporator is enclosed in the air distribution ductwork and can only be observed by removing a panel or part of the furnace plenum. High pressure liquid refrigerant enters the evaporator and expands into a gas, absorbing heat from the surrounding air.

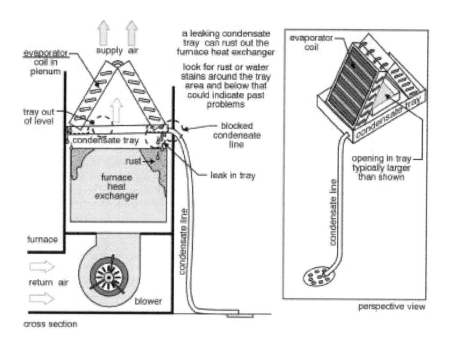

Leaking condensate tray

Air is pushed past the evaporator coil by the system's circulation blower; in the process, water vapor from the air condenses on the evaporator coil and drips into a drain pan. From there, it is directed to a condensate drain line that may sometimes include a condensate pump. The drain line empties into a house drain or directly to the building's exterior.

Examine the ductwork around the evaporator for signs of air leakage and check below the evaporator for signs of water leakage due to a blocked condensate drain line. Such leakage can present a serious problem if the evaporator is located above a warm air furnace, where dripping condensate water can rust the heat exchanger (see Section 7.3), or above a ceiling, where it can damage the building components below. Follow the condensate line and make sure that it terminates in a proper location. If there is a pump on the line, check its operation.

In split systems where the evaporator is located in an attic or closet, the condensate drain pan should have an auxiliary condensate drain line located above the regular drain line or an auxiliary drain pan that is separately drained. The connection of a condensate drain line to a plumbing vent in the attic may violate local codes. Check for such violations.

Test: If the evaporator coil can be exposed, inspect it for frost build up after about 30 minutes of operation. Frost is an indication of inadequate air flow due to dirt on the coil or a deficiency of refrigerant in the system. Check to see if water is discharging from the condensate drain line. If it is not, either the evaporator coil is not working properly or the drain line is clogged.

Test: Central air conditioning systems can be tested by an HVAC service technician to determine their overall condition and operational efficiency. This test requires a variety of specialized equipment and involves: 1) testing the pressure in the refrigerant lines, 2) taking amperage readings on the compressor, and 3) taking temperature readings of the air passing over the condenser and the evaporator coils, and correlating these readings with ambient outside temperature conditions.

- **Geothermal heating and cooling system.** Geothermal systems are relatively new and operate similarly to air-to-air heat pump systems, but differ in design and installation. What might be considered the condenser are pipes buried in the ground in dry wells or other in-ground systems suitable for transferring or displacing heat. The system is closed and its piping is PVC so corrosion is not a potential problem. Geothermal systems are normally installed without a back up or emergency heating system and all their components except the buried coils are usually inside the house. For more information, consult the *Geothermal Heat Pumps: Introductory Guide* published by the International Ground Source Heat Pump Association. A geothermal heating and cooling system can be operated in a heating or cooling mode under any outside temperature. Although expensive to install, they normally are efficient and economical to operate.

  Test: Observe the operation of the system in both heating and cooling modes if possible. Visually inspect the well field header piping and pumps. Pressure test the well field. Tests should be conducted by a service technician.

- **Distribution ductwork and controls.** Cool air distribution ductwork and controls, including zone controls, should be inspected similarly to those for forced warm air heating systems, as described in Section 7.3.

7.8 Central Gas-Absorption Cooling Systems

Gas-absorption cooling systems occasionally may be found in older residential buildings. Such systems use the evaporation of a liquid, such as ammonia, as the cooling agent and, like a gas refrigerator, are powered by a natural gas or propane flame.

Test: A gas-absorption system operates under several hundred pounds of pressure and should be tested by a specialist. The local gas or fuel supplier probably maintains the unit; ask for an evaluation of the system. Meanwhile, operate the system. It should start smoothly and run quietly.

Examine the condition of the system's exterior and interior components. Inspect ductwork in accordance with Section 7.3.

7.9 Heat Pumps

Electric heat pumps are electrically operated, refrigerant-type air conditioning systems that can be reversed to extract heat from outside air and transfer it indoors. Heat pumps are normally sized for their air conditioning load, which in most parts of the country is smaller than the heating load. Auxiliary electric heaters are used to provide the extra heating capacity the system requires in the heating season.

Like air conditioning systems, heat pumps can be either split or integral. Integral systems located outside the building should have well-insulated air ducts between the unit and the building. If located on or within the building, they should be mounted on vibration isolators, be thermally protected, and have an adequate condensate drainage system.

Inspect heat pumps by the procedure described in Section 7.7 for central air conditioning systems. Testing in one mode is usually sufficient. However, do not operate air-to-air heat pumps in temperatures below 65 °F (18 °C) on the cooling cycle and above 55 °F (13 °C) on the heating cycle. In both conditions the amount of work the heat pump has to do to achieve interior comfort temperatures is not enough to really test the heat pump's ability to perform. Check also for the following problems:

- **Auxiliary heater failure.** Electric resistance auxiliary heaters are designed to activate (usually in stages) below about 30 °F (-1 °C) outdoor temperature when the heat pump cannot produce enough heat to satisfy the thermostat. Examine the condition of all auxiliary heaters in accordance with Section 7.6.

  Test: If possible, activate the auxiliary heaters to observe their operation. Operating failures may be caused by a faulty heater element, faulty relays, a faulty thermostat, or a faulty reversing valve.

- **Improper defrosting.** During cold, damp weather, frost or ice may form on the metal fins of the coil in an outdoor unit. Heat pumps are designed to defrost this build up by reversing modes either at preset intervals or upon activation by a pressure sensing device.

Test: The pressure sensor may malfunction, and on units so equipped, it can be tested in the heating mode by temporarily placing an obstruction on the exhaust side of the coil and observing whether the coil begins to heat (defrost) within a short period of time.

■ **Faulty reversing valve.** In most heat pumps, a reversing valve changes modes from heating to cooling (some heat pumps use a series of dampers instead) when the thermostat is changed.

Test: Check the reversing valve when the outside temperature varies enough to be able to run the heat pump in the opposite modes. Change the thermostat. If the system doesn't reverse, the reversing valve is faulty.

7.10 Evaporative Cooling Systems

Evaporative cooling systems are simple and economical devices. They pass air through wetted pads or screens and cooling takes place by evaporation. Such systems can only be used in dry climates where evaporation readily

A roof-mounted evaporative cooler. This type of cooling unit is relatively simple to inspect, repair, and maintain.

Installation of this air conditioner destroyed the structural integrity of the bearing wall between the two windows.

takes place and where dehumidification is not required.

Evaporative coolers consist of evaporator pads or screens, a means to wet them, an air blower, and a water reservoir with a drain and float-operated water supply valve. These components are contained in a single housing, usually located on the roof, and connected to an interior air distribution system. In **wetted-pad coolers**, evaporator pads are wetted by a circulating pump that continually trickles water over them; in **slinger coolers**, evaporator pads are wetted by a spray; and in **rotary coolers**, evaporator screens are wetted by passing through a reservoir on a rotating drum.

The water in evaporative coolers often contains algae and bacteria that emit a characteristic "swampy" odor. These can be removed easily with bleach. Some systems counteract this pattern by treating the water or by continually adding a small amount of fresh water.

Inspect evaporative cooling systems by examining the condition of each component. Note whether evaporative pads need cleaning or replacement. Look for signs of leakage and check the cleanliness and operation of the water reservoir, float-operated supply valve, and drain.

Test: Activate the system and listen for unusual sounds or vibrations. Inspect all distribution ductwork as described in Section 7.3 and evaluate the system's overall ability to cool the building.

If the building is in seismic zones 3 or 4 (California and portions of Alaska, Arkansas, Hawaii, Idaho, Missouri, Montana, Nevada, Oregon, Utah, Wyoming, and Washington), check the evaporative cooler for the presence of seismic bracing to the structure.

7.11 Humidifiers

Humidifiers are sometimes added to warm air heating systems to reduce interior dryness during the heating season. They are installed with the air distribution system and are controlled by a humidistat that is usually located in the return air duct near the humidifier housing. The humidifier should not be located in a return air duct (see Section 7.3). Humidifiers can be of several types:

- **Stationary pad,** in which air is drawn from the furnace plenum or supply air duct by a fan, blown over an evaporator pad, and returned to the air distribution system.
- **Revolving drum,** in which water from a small reservoir is picked up by a revolving pad and exposed to an air stream from the furnace plenum or supply ductwork.
- **Atomizer,** in which water is broken into small particles by an atomizing device and released into the supply air ductwork.
- **Steam,** in which water is heated to temperatures above boiling and then injected into the supply air duct.

Inspect the humidifier's condition. Take off the unit's cover and check for mineral build up on the drum or pad. Examine the humidifier's water supply and look for signs of leakage, especially at its connection with the house water supply. Check all electrical wiring and connections.

Test: Activate the humidifier by turning up the humidistat. It should only operate when the furnace fan is on, the system is in the heating mode, and the indoor humidity is lower than the humidistat setting.

7.12 Unit (Window) Air Conditioners

Unit air conditioners are portable, integral air conditioning systems without ductwork. Inspect their overall condition and check the seal around each unit and its attachment to the window or wall. Ensure that it is adequately supported and look for obstructions to air flow on the exterior and for proper condensate drainage. Make sure all electrical service is properly sized and that each unit is properly grounded. Bent fins on the condenser coils may be "combed."

Test: Operate each window unit for a long enough period to determine its cooling capacity; after several minutes, the air from the unit should feel quite cool. It should start smoothly and run quietly. Check for water dripping from the condensate discharge on the exterior side of the unit.

7.13 Whole House and Attic Fans

Check the location and condition of the whole house or attic fan, if one is present (see Section 3.9). Inspect fan motors for signs of overheating and examine fan belts for signs of wear. Check all operating controls and associated electrical wiring and check to see that the attic fan thermostat is set at about 95 °F (35 °C).

Test: Activate whole house and attic fans and observe their operation. They should start and run smoothly and be securely fastened to their frames. Note whether the louvers below a whole house fan are open completely when the fan is running and whether exterior louvers on attic fans are weather protected and screened.

Appendix A— The Effects of Fire on Structural Systems

Introduction

Building fires, which normally reach temperatures of about 1000 °C, can affect the load-bearing capacity of structural elements in a number of ways. Apart from such obvious effects as charring and spalling, there can be a permanent loss of strength in the remaining material and thermal expansion may cause damage in parts of the building not directly affected by the fire.

In assessing fire's effects, the main emphasis should be placed on estimating the residual load-carrying capacity of the structure and then determining the remedial measures, if any, needed to restore the building to its original design for fire resistance and other requirements. Obviously, if weaknesses in the original design are exposed, these should be corrected.

All building materials except timber are likely to show significant loss of strength when heated above 250 °C, strength that may not recover after cooling. Thus, it is useful to estimate the maximum temperature attained in a fire. Molded glass objects soften or flow at 700 or 800 °C. Metals form drops or lose their sharp edges as follows: 300 to 350 °C for lead, 400 °C for zinc, 650 °C for aluminum and alloys, 950 °C for silver, 900 to 1000 °C for brass, 1000 °C for bronze, 1100 °C for copper, and 1100 to 1200 °C for cast iron. There are also the well-known color changes in concrete or mortar. The development of red or pink coloration in concrete or mortar containing natural sands or aggregates of appreciable iron oxide content occurs at 250 to 300 °C and, normally, 300 °C may be taken as the transition temperature. Table A-1 provides specifics.

Making an analysis of the damage and assessment of the necessary repairs may be possible within a reasonable degree of accuracy, but final acceptance may depend on proof by a load test, where performance is generally judged in terms of the recovery of deflection after load removal.

Table A–1
Fire-Induced Color Change in Concrete

The temperature within a slab may continue to rise after the fire has ended and some of the maxima were attained after the end of the heating period.

Heating period, hours	Maximum surface temperature		Maximum depth of concrete showing characteristic change			
	°F	°C	Pink or red 300 °C	Fading of red, friability 600 °C	Buff 600 °C	Sintering 1200 °C
1	1742	950	56 mm	19 mm	0	0
2	1922	1050	100 mm	38 mm	6 mm	0
4	2246	1230	140 mm	63 mm	25 mm	3 mm
6	2282	1250	170 mm	90 mm	38 mm	6 mm

1
Timber

Timber browns at about 120 to 150 °C, blackens around 200 to 250 °C, and emits combustible vapors at about 300 °C. Above a temperature of 400 to 450 °C (or 300 °C if a flame is present), the surface of the timber will ignite and char at a steady rate. Table A-2 shows the rate of charring.

Analysis and Repair

Generally, any wood that is not charred should be considered to have full strength. It may be possible to show by calculation that a timber section or structural element subjected to fire still has adequate strength once the char is removed. Where additional strength is required, it may be possible to add strengthening pieces. Joints that may have opened and metal connections that may have conducted heat to the interior are points of weakness that should be carefully examined.

2
Masonry

The physical properties and mechanisms of failure in masonry walls exposed to fire have never been analyzed in detail. Behavior is influenced by edge conditions and there is a loss of compressive strength as well as unequal thermal expansion of the two faces. For solid bricks, resistance to the effects of fire is directly proportional to thickness. Perforated bricks and hollow clay units are more sensitive to thermal shock. There can be cracking of the connecting webs and a tendency for the wythes to separate. In cavity walls, the inner wythe carries the major part of the load. Exterior walls can be subjected to more severe forces than internal walls by heated and expanding floor slabs. All types of brick give much better performance if plaster is applied, which improves insulation and reduces thermal shock.

Analysis and Repair

As with concrete, it is possible to determine the degree of heating of the wall from the color change of the mortar and bricks. For solid brick walls without undue distortion, the portion beyond the pink or red boundary may be considered serviceable and calculations should be made accordingly. Per-forated and hollow brick walls should be inspected for the effects of cracks indicating thermal shock. Plastered bricks sometimes suffer little damage and may need repairs only to the plaster surfaces.

3
Steel

The yield strength of steel is reduced to about half at 550 °C. At 1000 °C, the yield strength is 10 percent or less. Because of its high thermal conductivity, the temperature of unprotected internal steelwork normally will vary little from that of the fire. Structural steelwork is, therefore, usually insulated.

Table A–2
Char Rate of Timber

A column exposed to fire on all faces should be assumed to char equally on all faces 1.25 times faster than the rates shown. Linear interpolation or extrapolation for periods between 15 and 90 minutes is permissible.

Species	Charring after 30 minutes	Charring after 60 minutes
All structural species except those below	20 mm	40 mm
Western red cedar	25 mm	50 mm
Oak, utile, kerving (gurgun), teak, greenheart, jarrah	15 mm	30 mm

Apart from losing practically all of its load-bearing capacity, unprotected steelwork can undergo considerable expansion when sufficiently heated. The coefficient of expansion is 10^{-5} per degree Celsius. Young's modulus does not decrease with temperature as rapidly as does yield strength.

Cold-worked reinforced bars, when heated, lose their strength more rapidly than do hot-rolled high-yield bars and mild-steel bars. The differences in properties are even more important after heating. The original yield stress is almost completely recovered on cooling from a temperature of 500 to 600 °C for all bars but on cooling from 800 °C, it is reduced by 30 percent for cold-worked bars and by 5 percent for hot-rolled bars.

The loss of strength for prestressing steels occurs at lower temperatures than that for reinforcing bars. Cold-drawn and heat-treated steels lose a part of their strength permanently when heated to temperatures in excess of about 300 °C and 400 °C, respectively.

The creep rate of steel is sensitive to higher temperatures and becomes significant for mild steel above 450 °C and for prestressing steel above 300 °C. In fire resistance tests, the rate of temperature rise when the steel is reaching its critical temperature is fast enough to mask any effects of creep. When there is a long cooling period, however, as in prestressed concrete, subsequent creep may have some effect in an element that has not reached the critical condition.

Analysis and Repair

In general, a structural steel member remaining in place with negligible or minor distortions to the web, flanges, or end connections should be considered satisfactory for further service. Exceptions are the relatively small number of structures built with cold-worked or tempered steel, where there may be permanent loss of strength. This may be assessed using estimates of the maximum temperatures attained or by on-site testing. Where necessary, the steel should be replaced, although reinforcement with plates may be possible. Microscopy can be used to determine changes in microstructure. Since this is a specialized field, the services of a metallurgist are essential.

4
Concrete

Concrete's compressive strength varies not only with temperature but also with a number of other factors, including the rate of heating, the duration of heating, whether the specimen was loaded or not, the type and size of aggregate, the percentage of cement paste, and the water/cement ratio. In general, concrete heated by a building fire always loses some compressive strength and continues to lose it on cooling. However, where the temperature has not exceeded 300 °C, most strength eventually is recovered.

Because of the comparatively low thermal diffusivity of concrete (of the order of 1 mm/s), the 300 °C contour may be at only a small depth below the heated face. Concrete's modulus of elasticity also decreases with temperature, although it is believed that it will recover substantially with time, provided that the coefficient of thermal expansion of the concrete is on the order of 10^{-5} per degree Celsius (but this varies with aggregate). Creep becomes significant at quite low temperatures, being of the orders of 10^{-4} to 10^{-3} per hour over the temperature range of 250 to 700 °C, and can have a beneficial effect in relaxing stresses.

Analysis and Repair

- **Effective cross section.** Removal of the surface material down to the red boundary (see Table A-1) will reveal the remaining cross section that can be deemed effective. Compression tests of cores can indicate the strength of the concrete, yielding a value for use in calculations.

- **Cracks.** Most fine cracks are confined to the surface. Major cracks that could influence structural behavior are generally obvious. A wide crack or cracks near supports may mean there has been a loss of anchorage of the reinforcement.

- **Reinforcing steel.** Provided that mild steel or hot-rolled high-yield steel is undistorted and has not reached a temperature above about 800 °C, the steel may be assumed to have resumed its original properties except that cold-worked bars will have suffered some permanent loss.

- **Prestressing steel.** It is likely that prestressing steel will have lost some strength, particularly if it has reached temperatures over 400 °C. There will also be a loss of tensile stress. These effects can be assessed for the estimated maximum temperature attained.

In some situations, the replacement of a damaged concrete structural member may be the most practical and economic solution. Elsewhere, the repair of the member, even if extensive, will be justified to avoid inconvenience and damage to other structural members.

Where new members are connected to existing ones, monolithic action must be ensured. This calls for careful preparation of the concrete surfaces and the continuity of reinforcing steel. For repair, the removal of all loose friable concrete is essential to ensure adequate bonding. Extra reinforcement should be fastened only by experienced welders.

New concrete may be placed either by casting in forms or by the gunite method. With the latter, it may be possible to avoid increasing the original dimensions of the member. The choice of method will depend on the thickness of the new concrete, the surface finish required, the possibility of placing and compacting the concrete in the forms, and the degree of importance attached to an increase in the size of the section.

Large cracks can be sealed by injecting latex solutions, resins, or epoxies. Various washes or paints are available to restore the appearance of finely cracked or crazed surfaces.

Appendix B— Wood-Inhabiting Organisms

This material is excerpted from *A Guide to the Inspection of Existing Homes for Wood Inhabiting Insects*, by Michael P. Levy and published by the U.S. Department of Housing and Urban Development (HUD).

Wood is a porous material and will absorb moisture from the air. Moisture is attracted to the walls of the tubes that make up the wood. As walls absorb moisture, the wood swells. If the humidity is kept at 100 percent, the walls become saturated with water. The moisture content at which this occurs is the fiber saturation point, which is approximately 30 percent by weight for most species used in construction. Fungi will only decay wood with a moisture content above the fiber saturation point. To allow a safety margin, wood with a moisture content above 20 percent is considered to be susceptible to decay. Wood in properly constructed buildings seldom will have a moisture content above 16 to 18 percent. Thus, wood will only decay if it is in contact with the ground or wetted by an external source of moisture, such as rain seepage, plumbing leaks, or condensation. **Dry wood will never decay.** Also, the drier the wood, the less likely it is to be attacked by most types of wood-inhabiting insects.

Wood-inhabiting fungi are small plants that lack chlorophyll and use wood as their food source. Some fungi use only the starch and proteins in the wood and don't weaken it. Others use the structural components, and as they grow, they weaken the wood, which eventually becomes structurally useless. All fungi require moisture, oxygen, warmth, and food. The keys to preventing or controlling growth of fungi in wood in buildings are to either keep the wood dry (below a moisture content of 20 percent) or to use preservative-treated or naturally resistant heartwood or selected species.

Wood-inhabiting insects can be divided into those that use wood as a food material—termites and wood-boring beetles, for example—and those that use it for shelter—carpenter ants and bees, for example. Damage is caused by immature termites called nymphs, by the larvae or grubs of the wood-boring beetles, and by the adults in ants and bees.

Some wood-inhabiting organisms are found in all parts of the country, others are highly localized. Some, although common, cause very little structural damage. The following is a description of the major wood-inhabiting fungi and insects in the United States.

■ **Surface molds and sapstain fungi.** Surface molds or mildew fungi discolor the surface of wood, but do not weaken it. They are generally green, black, or orange and powdery in appearance. The various building codes allow the use of framing lumber with surface molds or mildew, providing that the wood is dry and not decayed. Spores (or seeds) of surface molds or mildew fungi grow quickly on moist wood or on wood in very humid conditions. They can grow on wood before it is seasoned, when it is in the supplier's yard or on the building site, or in a finished house. When the wood dries, the fungi die or become dormant, but they do not change their appearance. Thus, wherever surface molds or mildew fungi are observed on wood in a building, it is a warning sign that at some time the wood was moist or humidity was high.

Surface molds and mildew fungi are controlled by eliminating the source of high humidity or excess moisture, for example by repairing leaks, improving ventilation in attics or crawl spaces, or installing soil covers. Before taking corrective action, the source of the moisture that allowed fungus growth must be determined. If the wood is dry and the sources of moisture are no longer present, no corrective action need be taken.

Sapstain or bluestain fungi are similar to surface molds, except that the discoloration goes deep into the wood. They color the wood blue, black, or gray and do not weaken it. They grow quickly on moist wood and do not change their appearance when they die or become dormant. They usually occur in the living tree or before the wood is seasoned, but sometimes they grow in the supplier's yard, on the building site, or in a finished house. In the latter case, they are normally associated with rain seepage or leaks. Stain fungi are a warning sign that at some time the wood was moist. Control is the same as for surface molds or mildew fungi.

- **Water-conducting fungi.** Most decay fungi are able to grow only on moist wood and cannot attack adjacent dry wood. Two brown-rot fungi, *Poria incrassata* and *Merulius lacrymans,* are able to conduct water for several feet through root-like strands or rhizomorphs, to moisten wood and then to decay it. These are sometimes called water-conducting or dry-rot fungi. They can decay wood in houses very rapidly, but fortunately they are quite rare. *Poria incrassata* is found most frequently in the Southeast and West. *Merulius lacrymans* occurs in the Northeast. Both fungi can cause extensive damage in floors and walls away from obvious sources of moisture. Decayed wood has the characteristics of brown rotted wood except that the surface of the wood sometimes appears wavy but apparently sound, although the interior may be heavily decayed. The rhizomorphs that characterize these fungi can be up to an inch in diameter and white to black in color, depending on their age. They can penetrate foundation walls and often are hidden between wood members. The source of moisture supporting the fungal growth must be found and eliminated to control decay. Common sources include water leaks and wood in contact with or close to the soil: for example, next to earth-filled porches or planters. Where the fungus grows from a porch, the soil should be removed from the porch next to the foundation wall to prevent continued growth of the fungus into the house. *Poria incrassata* normally occurs in new or remodeled houses and can cause extensive damage within two to three years.

- **Brown-rot and white-rot fungi.** The fungi often produce a whitish, cottony growth on the surface of wood. They grow only on moist wood. The fungi can be present in the wood when it is brought into the house or can grow from the spores that are always present in the air and soil. Wood attacked by these fungi should not be used in construction.

  Wood decayed by brown-rot fungi is brittle and darkened in color. As decay proceeds, the wood shrinks, twists, and cracks perpendicular to the grain. Finally, it becomes dry and powdery. Brown-rot is the commonest type of decay found in wood in houses.

  Wood decayed by white-rot fungi is fibrous and spongy and is bleached in color. Sometimes it has thin, dark lines around decayed areas. The wood does not shrink until decay is advanced.

  These fungi can be controlled by eliminating the source of moisture that allows them to grow, for example by improving drainage and ventilation under a house, repairing water leaks, or preventing water seepage. When the wood dries, the fungi die or become dormant. Spraying wood with chemicals does not control decay. If the moisture source cannot be eliminated, all the decayed wood should be replaced with pressure-treated wood.

- **White-pocket rot.** White-pocket rot is caused by a fungus that attacks the heartwood of living trees. Decayed wood contains numerous small, spindle-shaped white pockets filled with fungus. These pockets are generally 3 to 13 mm long. When wood from infected trees is seasoned, the fungus dies. Therefore, no control is necessary. White-pocket rot generally is found in softwood lumber from the West Coast.

- **Subterranean termites.** Subterranean termites normally damage the interior of wood structures. Shelter tubes are the most common sign of their presence. Other signs include structural weakness of wood members, shed wings or warmers, soil in cracks or crevices, and dark or blister-like areas on wood. The major characteristics of infested softwood when it is broken open are that damage is normally greatest in the softer springwood and that gallery walls and inner surfaces of shelter tubes have a pale, spotted appearance like dried oatmeal. The galleries often contain a mixture of soil and digested wood. Termites usually enter houses through wood in contact with the soil or by building shelter tubes on foundation walls, piers, chimneys, plumbing, weeds, etc. Although they normally maintain contact with the soil, subterranean termites can survive when they are isolated from the soil if they have a continuing source of moisture. Heavy damage by subterranean termites (except Formosans) does not normally occur during the first five to 10 years of a building's life, although their attack may start as soon as it is built. Subterranean termites can be controlled most effectively by the use of chemicals in the soil and foundation area of the house, by breaking wood-soil contact, and by eliminating excess moisture in the house. When applied properly, these chemicals will prevent or control termite attack for at least 25 years.

- **Formosan subterranean termites.** Formosan subterranean termites are a particularly vigorous species of subterranean termite that has spread to this country from the Far East. They have caused considerable damage in Hawaii and Guam and have been found in several locations on the United States mainland. It is anticipated that they could eventually become established along southern coasts, the lower East and West Coasts, in the lower Mississippi Valley, and in the Caribbean.

The most obvious characteristics that distinguish Formosan subterranean termite swarmers from those of native species are their larger size (up to 16 mm compared to 9 to 13 mm) and hairy wings (compared with smooth wings in other subterraneans). Soldiers have oval shaped heads, as opposed to the oblong and rectangular heads of native soldiers. Formosan termites also produce a hard material called carton, which resembles sponge. This is sometimes found in cavities under fixtures or in walls adjacent to attacked wood. Other characteristics—and control methods—are similar to those for native subterranean termites. However, Formosan subterranean termites are more vigorous and can cause extensive damage more rapidly than do native species. For this reason Formosans should be controlled as soon as possible after discovery.

- **Drywood termites.** It is quite common for buildings to be infested by drywood termites within the first five years of their construction in southern California, southern Arizona, southern Florida, the Pacific area, and the Caribbean. Swarmers generally enter through attic vents or shingle roofs, but in hot, dry locations, they can be found in crawl spaces. Window sills and frames are other common entry points.

Drywood termites live in wood that is dry. They require no contact with the soil or with any other source of moisture. The first sign of drywood termite infestation is usually piles of fecal pellets, which are hard, less than 1 mm in length, with rounded ends and six flattened or depressed sides. The pellets vary in color from light gray to very dark brown, depending on the wood being consumed. The pellets, eliminated from galleries in the wood through round kick holes, accumulate on surfaces or in spider webs below the kick holes. There is very little external evidence of drywood termite attacks in wood other than the pellets. The interior of damaged wood has broad pockets or chambers that are connected by tunnels that cut across the grain through springwood and summerwood. The galleries are perfectly smooth and have few, if any, surface deposits. There are usually some fecal pellets stored in unused portions of the galleries. Swarming is another sign of termite presence.

It normally takes a very long time for the termites to cause serious weakness in house framing. Damage to furniture, trim, and hardwood floors can occur in a few years. The choice of control method depends on the extent of damage. If the infestation is widespread or inaccessible, the entire house should be fumigated. If infestation is limited, spot treatment can be used or the damaged wood can be removed.

- **Dampwood termites.** Dampwood termites of the desert Southwest and southern Florida are rarely of great danger to structures. Pacific

Coast dampwood termites can cause damage greater than subterranean termites if environmental conditions are ideal.

Dampwood termites build their colonies in damp, sometimes decaying wood. Once established, some species extend their activities to sound wood. They do not require contact with the ground, but do require wood with a high moisture content. There is little external evidence of the presence of dampwood termites other than swarmers or shed wings. They usually are associated with decayed wood. The appearance of wood damaged by dampwood termites depends on the amount of decay present. In comparatively sound wood, galleries follow the springwood. In decayed wood, galleries are larger and pass through both springwood and summerwood. Some are round in cross section, others are oval. The surfaces of the galleries have a velvety appearance and are sometimes covered with dried fecal material. Fecal pellets are about 1 mm long and colored according to the kind of wood being eaten. Found throughout the workings, the pellets are usually hard and round at both ends. In very damp wood, the pellets are often spherical or irregular, and may stick to the sides of the galleries.

Dampwood termites must maintain contact with damp wood. Therefore, they can be controlled by eliminating damp wood. Treatment of the soil with chemicals can also be used to advantage in some areas.

- **Carpenter ants.** Carpenter ants burrow into wood to make nests, but do not feed on the wood. They commonly nest in dead portions of standing trees, stumps, logs, and sometimes wood in houses. Normally they do not cause extensive structural damage. Most species start their nests in moist wood that has begun to decay. They attack both hardwoods and softwoods. The most obvious sign of infestation is the large reddish-brown to black ants, 6 to 13 mm long, inside the house. Damage occurs in the interior of the wood. There may be piles or scattered bits of wood powder (frass), which are very fibrous and sawdust-like. If the frass is from decayed wood, pieces tend to be darker and more square ended. The frass is expelled from cracks and crevices, or from slit-like openings made in the wood by the ants. It is often found in basements, dark closets, attics, under porches, and in crawl spaces. Galleries in the wood extend along the grain and around the annual rings. The softer springwood is removed first. The surfaces of the galleries are smooth, as if they had been sandpapered, and are clean. The most effective way to control carpenter ants is to locate the nest and kill the queen in colonies in and near the house with insecticides. It is sometimes also helpful to treat the voids in walls, etc. For current information on control, an entomologist should be contacted.

- **Wood-boring beetles, bees, and wasps.** There are numerous species of wood-boring insects that occur in houses. Some of these cause considerable damage if not controlled quickly. Others are of minor importance and attack only unseasoned wood. Beetles, bees, and wasps all have larval, or grub, stages in their life cycles, and the mature flying insects produce entry or exit holes in the surface of the wood. These holes, and sawdust from tunnels behind the holes, are generally the first evidence of attack that is visible to the building inspector. Correct identification of the insect responsible for the damage is essential if the appropriate control method is to be selected. The characteristics of each of the more common groups of beetles, bees, and wasps are discussed in the following table which summarizes the size and shape of entry or exit holes produced by wood-boring insects, the types of wood they attack, the appearance of frass or sawdust in insect tunnels, and the insect's ability to reinfest wood in a house.

To use the table, match the size and shape of the exit or entry holes in the wood to those listed in the table; note whether the damaged wood is a hardwood or softwood and whether damage is in a new or old wood product (evidence of inactive infestations of insects that attack only new wood will often be found in old wood; there is no need for control of these). Next, probe the wood to determine the appearance of the frass. It should then be possible to identify the insect type. It is clear from the table that there is often considerable variation within particular insect groups. Where the inspector is unsure of the identity of the insect causing damage, a qualified entomologist should be consulted.

- **Lyctid powder-post beetles.** Lyctids attack only the sapwood of hardwoods with large pores: for example, oak, hickory, ash, walnut, pecan, and many tropical hardwoods. They reinfest seasoned wood until it disintegrates. Lyctids range from 3 to 7 mm in length and are reddish-brown to black. The presence of small piles of fine flour-like wood powder (frass) on or under the wood is the most obvious sign of infestation.

Even a slight jarring of the wood makes the frass sift from the holes. There are no pellets. The exit holes are round and vary from 1 to 1.5 mm in diameter. Most of the tunnels are about 1.5 mm in diameter and loosely packed with fine frass. If damage is severe, the sapwood may be completely converted to frass within a few years and held in only by a very thin veneer of surface wood with beetle exit holes. The amount of damage depends on the level of starch in the wood. Infestations are normally limited to hardwood paneling, trim, furniture, and flooring. Replacement or removal and fumigation of infested materials are usually the most economical and effective control methods. For current information on the use of residual insecticides, the

Table B-1
Characteristics of Wood-Inhabiting Organisms

Shape and size (inches) of exit/entry hole	Wood type	Age of wood attacked	Appearance of frass in tunnels	Insect type	Reinfest
round, 0.5 to 1.5 mm	softwood & hardwood	new	none present	ambrosia beetles	no
round, 0.8 to 1.5 mm	hardwood	new & old	fine, flour-like, loosely packed	lyctid beetles	yes
round, 0.8 to 2.5 mm	bark/sapwood interface	new	fine to course, bark colored, tightly packed	bark beetles	no
round, 0.8 to 1.5 mm	softwood & hardwood	new & old	fine powder and pellets, loosely packed; pellets may be absent and frass tightly packed in some hardwoods	anobiid beetles	yes
round, 2.5 to 7 mm	softwood & hardwood (bamboo)	new	fine to course powder, tightly packed	bostrichid beetles	rarely
round, 1.5 to 7 mm	softwood	new	course, tightly packed	horntail or woodwasp	no
round, 13 mm	softwood	new & old	none present	carpenter beetle	yes
round-oval, 3 to 10 mm	softwood & hardwood	new	course to fibrous, mostly absent	round-headed borer	no
oval, 3 to 13 mm	softwood & hardwood	new	sawdust-like, tightly packed	flat-headed borer	no
oval, 6 to 10 mm	softwood	new & old	very fine powder and tiny pellets, tightly packed	old house borer	yes
flat oval 13 mm or more or irregular surface groove, 6 to 13 mm wide	softwood & hardwood	new	absent or sawdust-like, course to fibrous; tightly packed	round or flat headed borer	no

inspector should contact the extension entomologist at his nearest land grant university or a reputable pest control company.

- **Anobiid beetles.** The most common anobiids attack the sapwood of hardwoods and softwoods. They reinfest seasoned wood if environmental conditions are favorable. Attacks often start in poorly heated or ventilated crawl spaces and spread to other parts of the house. They rarely occur in houses on slab foundations. Anobiids range from 3 to 7 mm in length and are reddish-brown to nearly black. Adult insects are rarely seen. The most obvious sign of infestation is the accumulation of powdery frass and tiny pellets underneath infested wood or streaming from exit holes. The exit holes are round and vary from 1.5 to 3 mm in diameter. If there are large numbers of holes and the powder is bright and light colored like freshly sawed wood, the infestation is both old and active. If all the frass is yellowed and partially caked on the surface where it lies, the infestation has been controlled or has died out naturally. Anobiid tunnels are normally loosely packed with frass and pellets. It is normally 10 or more years before the number of beetles infesting wood becomes large enough for their presence to be noted. Control can be achieved by both chemical and non-chemical methods. For current information on control of anobiids, the inspector should contact the extension entomologist at his nearest land grant university or a reputable pest control company.

- **Bostrichid powderpost beetles.** Most bostrichids attack hardwoods, but a few species attack softwoods. They rarely attack and reinfest seasoned wood. Bostrichids range from 2.5 to 7 mm in length and from reddish-brown to black. The black polycaon is an atypical bostrichid and can be 13 to 25 mm in length. The first signs of infestation are circular entry holes for the egg tunnels made by the females. The exit holes made by adults are similar, but are usually filled with frass. The frass is meal-like and contains no pellets. It is tightly packed in the tunnels and does not sift out of the wood easily. The exit holes are round and vary from 2.5 to 9 mm in diameter. Bostrichid tunnels are round and range from 1.5 to 10 mm in diameter. If damage is extreme, the sapwood may be completely consumed. Bostrichids rarely cause significant damage in framing lumber and primarily affect individual pieces of hardwood flooring or trim. Replacement of structurally weakened members is usually the most economical and effective control method.

- **Old house borer.** This beetle infests the sapwood of softwoods, primarily pine. It reinfests seasoned wood, unless it is very dry. The old house borer probably ranks next to termites in the frequency with which it occurs in houses in the mid-Atlantic states. The beetle ranges from 15 to 25 mm in length, and is brownish-black in color. The first noticeable sign of infestation by the old house borer may be the sound of larvae boring in the wood. They make a rhythmic ticking or rasping sound, much like a mouse gnawing. In severe infestations the frass, which is packed loosely in tunnels, may cause the thin surface layer of the wood to bulge out, giving the wood a blistered look. When adults emerge (three to five years in the South, five to seven years in the North), small piles of frass may appear beneath or on top of infested wood. The exit holes are oval and 6 to 10 mm in diameter. They may be made through hardwood, plywood, wood siding, trim, sheetrock, paneling, or flooring. The frass is composed of very fine powder and tiny blunt-ended pellets. If damage is extreme, the sapwood may be completely reduced to powdery frass with a very thin layer of surface wood. The surfaces of the tunnels have a characteristic rippled pattern, like sand over which water has washed. Control can be achieved by both chemical and non-chemical methods. For current information on control of the old house borer, the inspector should contact the extension entomologist at his nearest land grant university or a reputable pest control company.

Appendix B—Wood-Inhabiting Organisms

■ **Carpenter bees.** Carpenter bees usually attack soft and easy-to-work woods, such as California redwood, cypress, cedar, and Douglas fir. Bare wood, such as unfinished siding or roof trim, is preferred. The only external evidence of attack is the entry holes made by the female. These are round and 9 mm in diameter. A rather course sawdust-like frass may accumulate on surfaces below the entry hole. The frass is usually the color of freshly sawed wood. The presence of carpenter bees in wood sometimes attracts woodpeckers, which increases the damage to the surface of the wood. The carpenter bee tunnels turn at a right angle after extending approximately an inch across the grain of the wood, except when entry is through the end of a board. They then follow the grain of the wood in a straight line, sometimes for several feet. The tunnels are smooth-walled. It takes several years of neglect for serious structural failure to occur. However, damaged wood is very unsightly, particularly if woodpeckers have followed the bees. The bees can be controlled by applying five to 10 percent carbaryl (Sevin) dust into the entry holes. Several days after treatment, the holes should be plugged with dowel or plastic wood. Prevention is best achieved by painting all exposed wood surfaces.

■ **Other wood-inhabiting insects.** There are several other species of insects that infest dying or freshly felled trees or unseasoned wood, but that do not reinfest seasoned wood. They may emerge from wood in a finished house or evidence of their presence may be observed. On rare occasions, control measures may be justified to prevent disfigurement of wood, but control is not needed to prevent structural weakening.

❏ *Ambrosia beetles.* These insects attack unseasoned sapwood and heartwood of softwood and hardwood logs, producing circular bore holes 0.5 to 3 mm in diameter. Bore holes do not contain frass, but are frequently stained blue, black, or brown. The insects do not infest seasoned wood.

❏ *Bark beetles.* These beetles tunnel at the wood/bark interface and etch the surface of wood immediately below the bark. Beetles left under bark edges on lumber may survive for a year or more as the wood dries. Some brown, gritty frass may fall from circular bore holes 1.5 to 2.5 mm in diameter in the bark. These insects do not infest wood.

❏ *Horntails (wood wasps).* Horntails generally attack unseasoned softwoods and do not reinfest seasoned wood. One species sometimes emerges in houses from hardwood firewood. Horntails occasionally emerge through paneling, siding, or sheetrock in new houses; it may take four to five years for them to emerge. They attack both sapwood and heartwood, producing a tunnel that is roughly C-shaped in the tree. Exit holes and tunnels are circular in cross-section and 1.5 to 7 mm in diameter. Tunnels are tightly packed with course frass. Frequently, tunnels are exposed on the surface of lumber by milling after the development of the insect.

❏ *Round-headed borers.* Several species are included in this group. They attack the sapwood of softwoods and hardwoods during storage, but rarely attack seasoned wood. The old house borer is the major round-headed borer that can reinfest seasoned wood. When round-headed borers emerge from wood, they make slightly oval to nearly round exit holes 3 to 10 mm in diameter. Frass varies from rather fine and meal-like in some species to very course fibers like pipe tobacco in others. Frass may be absent from tunnels, particularly where the wood was machined after the emergence of the insects.

❏ *Flat-headed borers.* These borers attack sapwood and heartwood of softwoods and hardwoods. Exit holes are oval, with the long diameter 3 to 13 mm. Wood damaged by flat-headed borers is generally sawed after damage has occurred, so tunnels are exposed on the surface of infested wood. Tunnels are packed with sawdust-like borings and pellets, and tunnel walls are covered with fine transverse lines somewhat similar to some round-headed borers. However, the tunnels are much more flattened. The golden buprestid is one species of flat-headed borer that occurs occasionally in the Rocky Mountain and Pacific Coast states. It produces an oval exit hole 5 to 7 mm across, and may not emerge from wood in houses for 10 or more years after infestation of the wood. It does not reinfest seasoned wood.

If signs of insect or fungus damage other than those already described are observed, the inspector should have the organism responsible identified before recommending corrective measures. Small samples of damaged wood, with any frass and insect specimens (larvae or grubs must be stored in vials filled with alcohol), should be sent for identification to the entomology or pathology department of the state land grant university.

Appendix C—Life Expectancy of Housing Components

The following material was developed for the National Association of Home Builders (NAHB) Economics Department based on a survey of manufacturers, trade associations, and product researchers. Many factors affect the life expectancy of housing components and need to be considered when making replacement decisions, including the quality of the components, the quality of their installation, their level of maintenance, weather and climatic conditions, and intensity of their use. Some components remain functional but become obsolete because of changing styles and tastes or because of product improvements. Note that the following life expectancy estimates are provided largely by the industries or manufacturers that make and sell the components listed.

Appliances

	Life in Years
Compactors	10
Dishwashers	10
Dryers	14
Disposal	10
Freezers, compact	12
Freezers, standard	16
Microwave ovens	11
Electric ranges	17
Gas ranges	19
Gas ovens	14
Refrigerators, compact	14
Refrigerators, standard	17
Washers, automatic and compact	13
Exhaust fans	20

Source: Appliance Statistical Review, April 1990

Bathrooms

	Life in Years
Cast iron bathtubs	50
Fiberglass bathtub and showers	10–15
Shower doors, average quality	25
Toilets	50

Sources: Neil Kelly Designers, Thompson House of Kitchens and Bath

Cabinetry

Kitchen cabinets	15–20
Medicine cabinets and bath vanities	20

Sources: Kitchen Cabinet Manufacturers Association, Neil Kelly Designers

Closet Systems

Closet shelves	Lifetime

Countertops

Laminate	10–15
Ceramic tile, high-grade installation	Lifetime
Wood/butcher block	20+
Granite	20+

Sources: AFPAssociates of Western Plastics, Ceramic Tile Institute of America

Doors

Screen	25–50
Interior, hollow core	Less than 30
Interior, solid core	30-lifetime
Exterior, protected overhang	80–100
Exterior, unprotected and exposed	25–30
Folding	30–lifetime
Garage doors	20–50
Garage door opener	10

Sources: Wayne Dalton Corporation, National Wood Window and Door Association, Raynor Garage Doors

Electrical

	Life in Years
Copper wiring, copper plated, copper clad aluminum, and bare copper	100+
Armored cable (BX)	Lifetime
Conduit	Lifetime

Source: Jesse Aronstein, Engineering Consultant

Finishes Used for Waterproofing

Paint, plaster, and stucco	3–5
Sealer, silicone, and waxes	1–5

Source: Brick Institute of America

Floors

Oak or pine	Lifetime
Slate flagstone	Lifetime
Vinyl sheet or tile	20–30
Terrazzo	Lifetime
Carpeting (depends on installation, amount of traffic, and quality of carpet)	11
Marble (depends on installation, thickness of marble, and amount of traffic)	Lifetime+

Sources: Carpet and Rug Institute, Congoleum Corporation, Hardwood Plywood Manufacturers Association, Marble Institute, National Terrazzo and Mosaic Association, National Wood Flooring Association, Resilient Floor Covering Institute

Footings and Foundation

Poured footings and foundations	200
Concrete block	100
Cement	50
Waterproofing, bituminous coating	10
Termite proofing (may have shorter life in damp climates)	5

Source: WR Grace and Company

Heating Ventilation and Air Conditioning

	Life in Years
Central air conditioning unit (newer units should last longer)	15
Window unit	10
Air conditioner compressor	15
Humidifier	8
Electric water heater	14
Gas water heater (depends on type of water heater lining and quality of water)	11–13
Forced air furnaces, heat pump	15
Rooftop air conditioners	15
Boilers, hot water or steam (depends on quality of water)	30
Furnaces, gas- or oil-fired	18
Unit heaters, gas or electric	13
Radiant heaters, electric	10
Radiant heaters, hot water or steam	25
Baseboard systems	20
Diffusers, grilles, and registers	27
Induction and fan coil units	20
Dampers	20
Centrifugal fans	25
Axial fans	20
Ventilating roof-mounted fans	20
DX, water, and steam coils	20
Electric coils	15
Heat Exchangers, shell-and-tube	24
Molded insulation	20
Pumps, sump and well	10
Burners	21

Sources: Air Conditioning and Refrigeration Institute, Air Conditioning, Heating, and Refrigeration News, Air Movement and Control Association, American Gas Association, American Society of Gas Engineers, American Society of Heating, Refrigeration and Air-Conditioning Engineers, Inc., Safe Aire Incorporated

Appendix C—Life Expectancy of Housing Components

Home Security Applicances

	Life in Years
Intrusion systems	14
Smoke detectors	12
Smoke/fire/intrusion systems	10

Insulation

For foundations, roofs, ceilings, walls, and floors	Lifetime

Sources: Insulation Contractors Association of America, North American Insulation Manufacturers Association

Landscaping

Wooden decks	15
Brick and concrete patios	24
Tennis courts	10
Concrete walks	24
Gravel walks	4
Asphalt driveways	10
Swimming pools	18
Sprinkler systems	12
Fences	12

Sources: Associated Landscape Contractors of America, Irrigation Association

Masonry

Chimney, fireplace, and brick veneer	Lifetime
Brick and stone walls	100+
Stucco	Lifetime

Sources: Brick Institute of America, Architectural Components, National Association of Brick Distributors, National Stone Association

Millwork

Stairs, trim	50–100
Disappearing stairs	30–40

Paints and Stains

	Life in Years
Exterior paint on wood, brick, and aluminum	7–10
Interior wall paint (depends on the acrylic content)	5–10
Interior trim and door paint	5–10
Wallpaper	7

Sources: Finnaren and Haley, Glidden Company, The Wall Paper

Plumbing

Waste piping, cast iron	75–100
Sinks, enamel steel	5–10
Sinks, enamel cast iron	25–30
Sinks, china	25–30
Faucets, low quality	13–15
Faucets, high quality	15–20

Sources: American Concrete Pipe Association, Cast Iron Soil and Pipe Institute, Neil Kelly Designers, Thompson House of Kitchens and Baths

Roofing

Asphalt and wood shingles and shakes	15–30
Tile (depends on quality of tile and climate)	50
Slate (depends on grade)	50–100
Sheet metal (depends on gauge of metal and quality of fastening and application)	20–50+
Built-up roofing, asphalt	12–25
Built-up roofing, coal and tar	12–30
Asphalt composition shingle	15–30
Asphalt overlag	25–35

Source: National Roofing Contractors Association

Rough Structure

	Life in Years
Basement floor systems	Lifetime
Framing, exterior and interior walls	Lifetime

Source: NAHB Research Foundation

Shutters

Wood, interior	Lifetime
Wood, exterior (depends on weather conditions)	4–5
Vinyl plastic, exterior	7–8
Aluminum, interior	35–50
Aluminum, exterior	3–5

Sources: A.C. Shutters, Inc., Alcoa Building Products, American Heritage Shutters

Siding

Gutters and downspouts	30
Siding, wood (depends on maintenance)	10–100
Siding, steel	50–Lifetime
Siding, aluminum	20–50
Siding, vinyl	50

Sources: Alcoa Building Products, Alside, Inc., Vinyl Siding Institute

Walls and Wall Treatments

Drywall and plaster	30–70
Ceramic tile, high grade installation	Lifetime

Sources: Association of Wall and Ceiling Industries International, Ceramic Tile Institute of America

Windows

Window glazing	20
Wood casement	20–50
Aluminum and vinyl casement	20–30
Screen	25–50

Sources: Best Built Products, Optimum Window Manufacturing, Safety Glazing Certification Council, Screen Manufacturers Association

Appendix D—References

The American Society of Home Inspectors (ASHI) is the professional organization that establishes home inspector qualifications and develops recommended home inspection standards. For information about this organization, write ASHI, Inc., 655 15th Street, N.W., Suite 320, Washington, D.C. 20005 (phone 202 842 3096), http://www.ashi.org.

Useful home inspection publications include:

Becker, Norman. 1993. *The Complete Book of Home Inspection*. New York: McGraw-Hill, Tab Books.

Boroson, Warren, and Ken Austin. 1993. *The Home Buyer's Inspection Guide*. New York: John Wiley & Sons.

Burgess, Russell W. 1999. *Real Estate Home Inspection*. 3d ed. Chicago, Illinois: Dearborn Financial Publishing Group, Real Estate Education Company.

Carson, Alan, and Robert Dunlop. 1999. *Inspecting a House: A Guide for Buyers, Owners, and Renovators*, 2d ed. New York: Stoddart Publishing Company.

Carson Dunlop & Associates Limited. 1998. *The Illustrated Home*. Toronto, Ontario: Carson Dunlop & Associates Limited. Available from Carson, Dunlop & Associates, 120 Carlton Street, Suite 407, Toronto, Ontario M5A4K2, Canada (phone 800 268 7070), http://www.carsondunlop.com

_____.1998. *Home Reference Book*. Toronto, Ontario: Carson Dunlop & Associates Limited. Available from Carson, Dunlop & Associates, 120 Carlton Street, Suite 407, Toronto, Ontario M5A4K2, Canada (phone 800 268 7070), http://www.carsondunlop.com

Irwin, Robert. 1995. *The Home Inspection Troubleshooter*. Chicago, IL: Dearborn Financial Publishing Group, Real Estate Education Company.

Scutella, Richard M., and Dave Heberle. 1994. *Home Buyer's Checklist, A Foolproof Guide to Finding the Perfect House*, 2d ed. New York: McGraw-Hill, Tab Books.

Traister, John E. 1997. *Home Inspection Handbook*. Gordon and Breach Publishing Group, Craftsman House Books.

Ventolo, William L. 1995. *Your Inspection Guide*. Chicago, IL: Dearborn Financial Publishing Group, Real Estate Education Company.

The CABO *One and Two Family Dwelling Code* and its successor, the *International Residential Code for One- and Two-Family Dwellings*, are the most widely used and accepted residential building codes in the United States and can be used for comparing conditions in existing structures against current code requirements. They also provide span and working stress tables for wood joists and rafters, which are useful for checking the adequacy of wood structural components.

International Code Council, Inc. 2000. *International Residential Code for One- and Two-Family Dwellings*. Falls Church, VA: International Code Council, Inc. Available from any of the three model code organizations: Building Officials and Code Administrators International, Inc., 4051 West Flossmoor Road, Country Club Hills, IL 60478-5795 (phone 703 799 2300), http://www.bocai.org; Southern Building Code Congress International, Inc., 900 Montclair Road, Birmingham, AL 35213-1206 (phone 205 591 1853), http://www.sbcci.org; and International Conference of Building Officials, 5360 Workman Mill Road, Whittier, CA 90601-2298 (phone 800 423 6587), http://www.icbo.org.

The *National Electrical Code* should be used as a reference for all electrical work.

National Fire Protection Association (NFPA). 2000. *National Electrical Code*. Quincy, MA: National Fire Protection Association. Available from NFPA, Batterymarch Park, Quincy, MA 02269 (phone 617 770 3000), http://www.nfpa.org.

The following publications are useful general references for rehabilitating houses:

Ching, Francis D. K., and Miller, Dale E. 1983. *Home Renovation*. New York: John Wiley & Sons.

Litchfield, Michael W. 1997. *Renovation: A Complete Guide*. New York: Sterling Publishing.

Nash, George. 1996. *Renovating Old Houses*. Newtown, CT: Taunton Press.

Vila, Bob, and Hugh Howard. 1999. *Bob Vila's Complete Guide to Remodeling Your Home: Everything You Need to Know about Home Renovation from the Number One Home Improvement Expert*. New York: Harper Collins, Avon Books.

Residential buildings of historic significance should be rehabilitated in accordance with the following:

Heritage Preservation Services, National Park Service (NPS). 2000. *The Secretary of the Interior's Standards for Rehabilitation and Illustrated Guidelines for Rehabilitating Historic Buildings*. Washington, DC: National Park Service. Available from NPS, Washington, DC 20240, http://www.nps.gov or full text online at http://www2.cr.nps.gov/tps/tax/rhb/stand.htm.

Publications useful for assessing the energy efficiency of existing residential buildings include:

U.S. Department of Housing and Urban Development (HUD). 1982. *Applying Cost-Effective Energy Standards in Rehabilitation Projects*. Washington, DC: U.S. Department of Housing and Urban Development. Available from HUD User, P.O. Box 6091, Rockville, MD 20849 (phone 800 245 2691), http://www.huduser.org.

_____. 1981. *Conserving Energy in Older Homes: A Do-It-Yourself Manual*. Washington, DC: U.S. Department of Housing and Urban Development. Available from HUD User, P.O. Box 6091, Rockville, MD 20849 (phone 800 245 2691), http://www.huduser.org.

_____. 1982. *Energy Conserving Features in Older Homes*. Washington, DC: U.S. Department of Housing and Urban Development. Available from HUD User, P.O. Box 6091, Rockville, MD 20849 (phone 800 245 2691), http://www.huduser.org.

_____. 1977. *In the Bank or Up the Chimney: A Dollars and Cents Guide to Energy-Savings Home Improvement*. Washington, DC: U.S. Department of Housing and Urban Development. Available from HUD User, P.O. Box 6091, Rockville, MD 20849 (phone 800 245 2691), http://www.huduser.org.

_____. 1996. *Rehabilitation Energy Guidelines for One-to-Four Family Dwellings*. Washington, DC: U.S. Department of Housing and Urban Development. Available from HUD User, P.O. Box 6091, Rockville, MD 20849 (phone 800 245 2691), http://www.huduser.org.

Wilson, Alex and John Morrill. 2000. *Consumer Guide to Home Energy Savings*. 6th ed. Washington, DC: American Council for an Energy Efficient Economy.

The following publications are referenced in this Guideline:

Air Conditioning Contractors of America (ACCA). *Residential Load Calculation* (Manual J). Washington, DC: Air Conditioning Contractors of America. Available from ACCA, 1712 New Hampshire Avenue, NW, Washington, DC 20009 (phone 202 483 9370), http://www.acca.org.

Ambrose, James E. 2000. *Simplified Engineering for Architects and Builders*. New York: John Wiley & Sons.

Federal Emergency Management Agency (FEMA). 1999. *Taking Shelter from the Storm: Building a Safe Room Inside Your House* (FEMA 320). 2d ed. Washington, DC: Federal Emergency Management Agency. Available from FEMA Publications, P.O. Box 2012, Jessup, MD 20794-2012 (phone 800 565 3896), http://www.fema.gov.

_____. 1996. *Design Manual for Retrofitting Flood-Prone Residential Structures* (FEMA-114). Washington, DC: Federal Emergency Management Agency. Available from FEMA Publications, P.O. Box 2012, Jessup, MD 20794-2012 (phone 800 565 3896), http://www.fema.gov.

Appendix D—References

Gypsum Association. 1999. *Design Data—Gypsum Products* (GA-530). Washington, DC: Gypsum Association. Available from Gypsum Association, 810 First Street, NW, Washington, DC 20002 (phone 202 289 5440), http://www.gypsum.org.

Institute for Business and Home Safety (IBHS). 1998. *Is Your Home Protected from Hurricane Disaster? A Homeowner's Guide to Hurricane Retrofit.* Boston: Institute for Business and Home Safety. Available from IBHS, 175 Federal Street, Suite 500, Boston, MA 02110-2222 (phone 617 292 2003), http://www.ibhs.org.

_____. 1999. *Is Your Home Protected from Earthquake Disaster? A Homeowner's Guide to Earthquake Retrofit.* Boston: Institute for Business and Home Safety. Available from IBHS, 175 Federal Street, Suite 500, Boston, MA 02110-2222 (phone 617 292 2003), http:// www.ibhs.org.

_____. 1999. *Is Your Home Protected from Hail Damage? A Homeowner's Guide to Roofing and Hail.* Boston: Institute for Business and Home Safety. Available from IBHS, 175 Federal Street, Suite 500, Boston, MA 02110-2222 (phone 617 292 2003),http://www.ibhs.org.

International Conference of Building Officials (ICBO). 1997. *Seismic Strengthening Provisions for Unreinforced Masonry Bearing Wall Buildings.* Appendix Chap. 1 in Uniform Code for Building Conservation. Whittier, CA: International Conference of Building Officials. Available from ICBO, 5360 Workman Mill Road, Whittier, CA 90601-2298 (phone 800 423 6587), http:// www.icbo.org.

International Ground Source Heat Pump Association (IGSHPA). 1997. *Geothermal Heat Pumps: Introductory Guide.* Stillwater, OK: International Ground Source Heat Pump Association. Available from IGSHPA, 490 Cordell South, Oklahoma State University, Stillwater, OK 74078-8018 (phone 800 626 4747), http:// www.igshpa.okstate.edu

Lightning Protection Institute (LPI). *Installation Code* (LPI-175). Arlington Heights, IL: Lightning Protection Institute. Available from LPI, 3335 North Arlington Heights Road, Suite E, Arlington Heights, IL 60004 (phone 847 577 7200), http://www.lightning.org.

National Association of Home Builders Remodelers Council and Single Family Small Volume Builders Committee (NAHB). 1996. *Residential Construction Performance Guidelines for Professional Builders and Remodelers.* Washington, DC: National Association of Home Builders. Available from NAHB, 1201 15th Street, NW, Washington, DC 20005 (phone 202 822 0200), http://www.nahb.org.

National Concrete Masonry Association (NCMA). 1997. *Sound Transmission Class Ratings for Concrete Masonry Walls* (Tek Note 13-1). Herndon, VA: National Concrete Masonry Association. Available from NCMA, 2302 Horse Pen Road, Herndon, VA 20171 (phone 703 713 1900), http:// www.ncma.org.

National Fire Protection Association (NFPA). 1997. *Standard for the Installation of Lightning Protection Systems* (NFPA 780). Quincy, MA: National Fire Protection Association. Available from NFPA, Batterymarch Park, Quincy, MA 02269 (phone 617 770 3000), http:// www.nfpa.org.

National Institute of Building Sciences (NIBS). 1996. *Guidance Manual: Asbestos Operations and Maintenance Work Practices.* Washington, DC: National Institute of Building Sciences. Available from NIBS, 1090 Vermont Avenue, NW, Washington, DC 20005-4905 (phone 202 289 7800), http:// www.nibs.org.

_____. 1995. *Lead-Based Paint Operation and Maintenance Work Practices Manual for Homes and Buildings.* Washington, DC: National Institute of Building Sciences. Available from NIBS, 1090 Vermont Avenue, NW, Washington, DC 20005-4905 (phone 202 289 7800), http://www.nibs.org.

Partnership for Advancing Technology in Housing (PATH). 1998–. *The Rehab Guide.* Vol. 1, *Foundations* (HUD 8590), vol. 2, *Exterior Walls* (HUD 8751), vol. 3, *Roofs* (HUD 8702), vol. 4, *Windows and Doors* (HUD 8724), vol. 5, *Partitions, Ceilings, Floors and Stairs* (HUD 8798), vol. 6, *Kitchens and Baths* (HUD 8796), vol. 7, *Electrical/Electronics*, vol. 8, *HVAC/Plumbing*

(HUD 8797), vol. 9, *Site Work*. Washington, DC: U.S. Department of Housing and Urban Development (HUD). Available from HUD User, P.O. Box 6091, Rockville, MD 20848 (phone 800 245 2691), http://www.huduser.org or full text online at Path Net, http://www.pathnet.org.

Underwriters Laboratories (UL). 1994. *Installation Requirements for Lightning Protection Systems* (UL 96A). Northbrook, IL: Underwriters Laboratories. Available from UL, 333 Pfingsten Road, Northbrook, IL 60062-2096 (phone 847 272 8800), www.ulstandardsinfonet.ul.com.

———. 1994. UL *Standard for Safety for Lightning Protection Components* (UL 96). Northbrook, IL: Underwriters Laboratories. Available from UL, 333 Pfingsten Road, Northbrook, IL 60062-2096 (phone 847 272 8800), http://www.ulstandardsinfonet.ul.com.

U. S. Department of Housing and Urban Development (HUD). 1996. *Residential Remodeling and Universal Design: Making Homes More Comfortable and Accessible* (HUD 07107). Washington, DC: U.S. Department of Housing and Urban Development. Available from HUD User, P.O. Box 6091, Rockville, MD 20848 (phone 800 245 2691), http://www.huduser.org.

———. 1995. Lead-Based Paint Inspections. Chap. 7 in *Guidelines for the Evaluation and Control of Lead-Based Paint Hazards in Housing*. Washington, DC: U.S. Department of Housing and Urban Development. Available from HUD User, P.O. Box 6091, Rockville, MD 20848 (phone 800 245 2691), http://www.huduser.org.

———. 1997. *Nationally Applicable Recommended Rehabilitation Provisions*. Washington, DC: U.S. Department of Housing and Urban Development. Available from HUD User, P.O. Box 6091, Rockville, MD 20848 (phone 800 245 2691), http://www.huduser.org.

The following test standards are referenced in this Guideline:

American Concrete Institute (ACI). 1992. *Guide for Making a Condition Survey of Concrete in Service* (ACI 201.1R-92). Farmington, MI: American Concrete Institute. Available from ACI, P.O. Box 9094, Farmington, MI 48333, http://www.aciint.org.

———. 1994. *Guide for Evaluation of Concrete Structures Prior to Rehabilitation* (ACI 364.1R-94). Farmington, MI: American Concrete Institute. Available from ACI, P.O. Box 9094, Farmington, MI 48333, http://www.aciint.org.

———. 1997. *Standard Specification for Repairing Concrete with Epoxy Mortars* (ACI 503.4-92). In Four Epoxy Standards (ACI 503.1-92). Farmington, Michigan: American Concrete Institute. Available from ACI, P.O. Box 9094, Farmington, MI 48333, http://www.aciint.org.

American Society for Testing and Materials (ASTM). 1997. *Standard Test Method for Penetration Resistance of Hardened Concrete* (ASTM C803/X803M-97e1). West Conshohocken, PA: American Society for Testing and Materials. Available from ASTM, 100 Barr Harbor Drive, West Conshohocken, PA 19428 (phone 610 832 9500), http://www.astm.org.

———. 1997. *Standard Test Method for Rebound Number of Hardened Concrete* (ASTM C805-97). West Conshohocken, PA: American Society for Testing and Materials. Available from ASTM, 100 Barr Harbor Drive, West Conshohocken, PA 19428 (phone 610 832 9500), http://www.astm.org.

———. 1999. *Standard Test Method for Effect of Air Supply on Smoke Density in Flue Gasses from Burning Distillate Fuels* (ASTM D2157-94 (1999)). West Conshohocken, PA: American Society for Testing and Materials. Available from ASTM, 100 Barr Harbor Drive, West Conshohocken, PA 19428 (phone 610 832 9500), http://www.astm.org.

———. 1999. *Standard Test Method for Tension Testing of Metallic Material* (ASTM E8-99). West Conshohocken, PA: American Society for Testing and Materials. Available from ASTM, 100 Barr Harbor Drive, West Conshohocken, PA 19428 (phone 610 832 9500), http://www.astm.org.

———. 1989. *Standard Test Method of Compression Testing of Metallic Materials at Room Temperature* (ASTM E9-89ae1). West Conshohocken, PA: American Society for Testing and Materials. Available from ASTM, 100 Barr Harbor Drive, West Conshohocken, PA 19428 (phone 610 832 9500), http://www.astm.org.

———. 1999. *Standard Method for Laboratory Measurement of Airborne Sound Transmission Loss of Building Partitions and Elements.* (ASTM E90-99). West Conshohocken, PA: American Society for Testing and Materials. Available from ASTM, 100 Barr Harbor Drive, West Conshohocken, PA 19428 (phone 610 832 9500), http://www.astm.org.

———. 1996. *Standard Method of Laboratory Measurement of Impact Sound Transmission Through Floor-Ceiling Assemblies Using the Tapping Machine* (ASTM E492-90(1996)e1). West Conshohocken, PA: American Society for Testing and Materials. Available from ASTM, 100 Barr Harbor Drive, West Conshohocken, PA 19428 (phone 610 832 9500), http://www.astm.org.

———. 1999. *Standard Test Methods for Flexural Bond Strength of Masonry* (ASTM E518-99). West Conshohocken, PA: American Society for Testing and Materials. Available from ASTM, 100 Barr Harbor Drive, West Conshohocken, PA 19428 (phone 610 832 9500), http://www.astm.org.

———. 1993. *Standard Test Method for Diagonal Tension (Shear) in Masonry Assemblages* (ASTM E519- 81(1993)e1). West Conshohocken, PA: American Society for Testing and Materials. Available from ASTM, 100 Barr Harbor Drive, West Conshohocken, PA 19428 (phone 610 832 9500), http://www.astm.org.

———. 1995. *Standard Test Method for Determining Air Change in a Single Zone by Means of a Tracer Gas Dilution* (ASTM E741-95). West Conshohocken, PA: American Society for Testing and Materials. Available from ASTM, 100 Barr Harbor Drive, West Conshohocken, PA 19428 (phone 610 832 9500), http://www.astm.org.

———. 1999. *Standard Test Method for Determining Air Leakage Rate by Fan Pressurization* (ASTM E779-99). West Conshohocken, PA: American Society for Testing and Materials. Available from ASTM, 100 Barr Harbor Drive, West Conshohocken, PA 19428 (phone 610 832 9500), http://www.astm.org.

———. 1993. *Standard Test Method for Field Measurement of Air Leakage Through Installed Exterior Windows and Doors* (ASTM E783-93). West Conshohocken, PA: American Society for Testing and Materials. Available from ASTM, 100 Barr Harbor Drive, West Conshohocken, PA 19428 (phone 610 832 9500), http://www.astm.org.

———. 1996. *Standard Test Method for Field Determination of Water Penetration of Installed Exterior Windows, Curtain Walls, and Doors by Uniform or Cyclic Static Air Pressure Difference* (ASTM E1105-96). West Conshohocken, PA: American Society for Testing and Materials. Available from ASTM, 100 Barr Harbor Drive, West Conshohocken, PA 19428 (phone 610 832 9500), http://www.astm.org.

American Society of Civil Engineers (ASCE). 1990. *Guideline for Structural Condition Assessment of Existing Buildings* (ASCE 11-90). Reston, VA: American Society of Civil Engineers. Available from ASCE, 1801 Alexander Bell Drive, Reston, VA 20191-4400 (phone 703 295 6300), http://www.asce.org.

Associated Air Balance Council (AABC). 1999. *AABC Test and Balance Procedures* (MN-4). Washington, DC: Associated Air Balance Council. Available from AABC, 1518 K Street, NW, Washington, DC 20005 (phone 202 737 0202), http://www.aabchq.com.

National Environmental Balancing Bureau (NEBB). 1998. *Procedural Standards for Testing, Adjusting, and Balancing of Environmental Systems*, 6th ed. Gaithersburg, MD: National Environmental Balancing Bureau. Available from NEBB, 8575 Grovemont Circle, Gaithersburg, MD 20877-4121 (phone 301 977 3698) http://www/nebb.org.

Appendix E—
Inspection Record

Building Name/Location

_____

_____

_____

Lot size _____

Year building built _____

Total building area _____

Applicable building codes _____

Setback requirements:

    Front yard _____

    Side yard _____

    Rear yard _____

Other _____

Zoning classification _____

Date of alterations _____

Height limitation _____

_____

_____

_____

_____

General comments:

_____

_____

_____

_____

_____

_____

_____

_____

_____

_____

_____

Site Plan

Note the following on the site plan:

- ❏ North arrow
- ❏ Lot lines
- ❏ Building outline
- ❏ Drainage direction(s)
- ❏ Outbuildings
- ❏ Sidewalks and driveways
- ❏ Plantings
- ❏ Fences and walls
- ❏ Utility lines
- ❏ Water well, if any
- ❏ Septic system, if any
- ❏ Drawing scale

Provide dimensions for all major site components

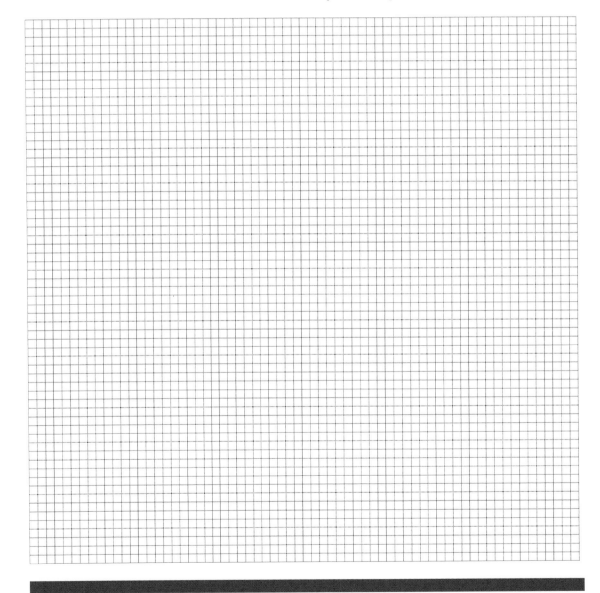

Elevations

Note the following on each elevation:

- ❏ All exterior doors and windows
- ❏ Important architectural details
- ❏ Floor-to-floor heights
- ❏ Material types
- ❏ Direction of view
- ❏ Drawing scale

Supplement with exterior photographs as appropriate

Floor plan

Note the following on each floor plan:

- ❏ North arrow
- ❏ Floor level
- ❏ Wall thicknesses
- ❏ Wall materials
- ❏ Exterior dimensions
- ❏ Window sizes
- ❏ Door widths and swings
- ❏ Room dimensions
- ❏ Plumbing fixtures
- ❏ HVAC equipment
- ❏ Kitchen cabinetry
- ❏ Scale

Show major structural elements in colored pencil or marker

Inspection checklist

1—Site

| Data | Condition/needed repairs |

1.1 Drainage
- Window well sizes
- Basement stairwell size

1.2 Site improvements
- Types of plantings
- Fence dimensions
- Lighting types
- Driveway dimensions
- Sidewalk widths
- Step dimensions
- Retaining walls

1.3 Outbuildings
- Garage dimensions
- Shed dimensions
- Other

1.4 Yards and Courts
- Areaway dimensions
- Lighting dimensions
- Access

1.5 Flood Region
- ❏ Flood risk zone (see local authorities)

2—Building Exterior

Data · Condition/needed repairs

2.1 Foundation Walls and Piers
- See Sections 4.1 and 4.2 for masonry
- See Section 4.7 for concrete
- See Section 4.5 for wood

2.2 Exterior Wall Cladding
- Cladding material
- Thermal insulation

2.3 Windows and Doors
- Door types — No.
- Window types — No.
- Storm window type — No.
- Storm door type — No.

2.4 Decks, Porches, Balconies
- Size(s)
- Flooring material(s)
- Railing height(s)

2.5 Pitched roofs ❑ Replace ❑ Retain
- Covering type
- Flashing type

2.6 Flat Roofs ❑ Replace ❑ Retain
- Covering type
- Flashing type

2.7 Skylights
- Size(s)

Data   Condition/needed repairs

2.8 Gutters, Downspouts and Drains

❏ Replace   ❏ Retain

Gutter size(s)
(1 sq. in. per 100 sq. ft. of roof)

Downspout size(s)
(one downspout per 40 ft. of gutter)

2.9 Chimneys

Height above roof

Flue size(s)

2.10 Parapets and Gables

❏ Requires structural inspection

2.11 Lightning Protection

❏ Protection required

3—Building Interior

	Data	Condition/needed repairs
3.1 Basement/Crawl Space		
Floor height		
Floor material		
Wall material		
Insulating materials		

3.2 Interior Spaces

	Data		Condition/needed repairs
Room			
Dimensions	Height		
Ceiling/wall material(s)			
Floor material			
Door size(s)			
Window size(s)			
Closet size(s)			
Trim			
No. 120V outlets	❑ 240V outlet		
Heat source			
Skylights			

	Data		Condition/needed repairs
Room			
Dimensions	Height		
Ceiling/wall material(s)			
Floor material			
Door size(s)			
Window size(s)			
Closet size(s)			
Trim			
No. 120V outlets	❑ 240V outlet		
Heat source			
Skylights			

Appendix E—Inspection Record

                Data   Condition/needed repairs

Room

Dimensions   Height

Ceiling/wall material(s)

Floor material

Door size(s)

Window size(s)

Closet size(s)

Trim

No. 120V outlets   ❏ 240V outlet

Heat source

Skylights

Room

Dimensions   Height

Ceiling/wall material(s)

Floor material

Door size(s)

Window size(s)

Closet size(s)

Trim

No. 120V outlets   ❏ 240V outlet

Heat source

Skylights

	Data	Condition/needed repairs

Room

Dimensions   Height

Ceiling/wall material(s)

Floor material

Door size(s)

Window size(s)

Closet size(s)

Trim

No. 120V outlets   ❏ 240V outlet

Heat source

Skylights

3.3 Bathroom

Dimensions   Height

Ceiling/wall material(s)

Floor/wall material(s)

Window size   Height from floor

Closet size(s)

Heat source

❏ 120V outlet   ❏ GFCI protected

Lavatory: ❏ Replace   ❏ Retain

Toilet: ❏ Replace   ❏ Retain

Tub/shower: ❏ Replace   ❏ Retain

Ventilation source

	Data		Condition/needed repairs
Dimensions	Height		
Ceiling/wall material(s)			
Floor/wall material(s)			
Window size	Height from floor		
Closet size(s)			
Heat source			
❑ 120V outlet	❑ GFCI protected		
Lavatory:	❑ Replace	❑ Retain	
Toilet:	❑ Replace	❑ Retain	
Tub/shower:	❑ Replace	❑ Retain	
Ventilation source			

3.4 Kitchen

	Data		Condition/needed repairs
Dimensions	Height		
Ceiling/wall material			
Floor covering			
Window size(s)			
Counter space, l.f.			
Overhead cabinets, l.f.			
Undercounter cabinets, l.f.			
Heat source			
No. of 120v outlets			
❑ Sep. 120V 20 amp refrig. outlet			
❑ 240V range outlet			
❑ gas outlet			
Dishwasher (20 amp.)	❑ Replace	❑ Retain	
Disposal (20 amp.)	❑ Replace	❑ Retain	
Exhaust fan:	❑ Replace	❑ Retain	
Other:	❑ Replace	❑ Retain	

	Data	Condition/needed repairs

3.5 Storage Spaces

Location	Size	
Location	Size	
Location	Size	

3.6 Stairs/Hallway

- Ceiling/wall material
- Floor material
- ❏ Three-way light control
- ❏ Smoke detector
- Handrail ht.   Railing ht.
- Tread/riser dim.
- Stair width   Head room
- Structural integrity

3.7 Laundry/Utility Room

- Ceiling/wall material
- Floor covering
- ❏ Plumbing connections adequate
- ❏ Dryer vent
- Laundry tub: ❏ Replace   ❏ Retain
- ❏ Floor drain present
- Washer: ❏ Replace   ❏ Retain
- Dryer: ❏ Replace   ❏ Retain
- ❏ 240V. outlet
- ❏ Gas outlet

3.8 Fireplace/flues

Opening	Location	Size	Depth

	Data	Condition/needed repairs

3.9 Attic

Ht. of highest point	
Means of access	
Ventilation, clear area ___ s.f.	
Signs of roof leakage	
Type of insulation ___ Depth	

3.10 Whole Building Thermal Efficiency Tests

- ❏ Conduct pressurization test
- ❏ Scan exterior for heat loss

3.11 Sound Transmission Control

- ❏ Conduct sound transmission tests

3.12 Asbestos

- ❏ Evidence of asbestos

3.13 Lead

- ❏ Evidence of lead-based paint
- ❏ Water test conducted

3.14 Radon

- ❏ Radon test conducted

3.15 Tornado Room

- ❏ Requires structural inspection

4—Structural System

	Data	Condition/needed repairs

4.1 Seismic Resistance
 ❏ Requires structural inspection

4.2 Wind Resistance
 ❏ Requires structural inspection

4.3 Masonry, General
 Load bearing walls are:

4.4 Masonry Foundations and Piers
 Foundation wall material
 Wall thickness
 Pier material
 Pier size(s)
 Pier spacing
 Depth of footings
 ❏ Structural problems

4.5 Above-ground Masonry Walls
 Wall material(s)
 Wall thickness
 Support over openings
 ❏ Thermal moisture cracking
 ❏ Freeze/thaw, corrosion cracking
 ❏ Structural failure cracking
 ❏ Wall bulging
 ❏ Wall leaning
 ❏ Brick veneer problems
 ❏ Parapet wall problems
 ❏ Fire damage problems

Appendix E—Inspection Record · E-15

                                     Data   Condition/needed repairs

4.6 Chimneys

    Chimney materials _____

    Depth of footings _____

    ❑ Structural problems _____

4.7 Wood Structural Components

    Framing type
    (balloon, platform, timber frame)

    Floor members   size   spacing

    Floor substrate material

    Wall members   size   spacing

    Wall substrate material

    Ceiling members   size   spacing

    Roof members   size   spacing

    Roof substrate material

    ❑ Deflection/warping problems

    ❑ Signs of fungal/insect attack

    ❑ Fire damage problems

4.8 Iron and Steel Structural Components

    Lintels, Columns and Beams

        size   location

        size   location

        size   location

    ❑ Lintel problems

    ❑ Column/beam problems

    ❑ Fire damage problems

	Data	Condition/needed repairs

4.9 Concrete Structural Components

Slabs, Lintels, Walls

    size   location

    size   location

    size   location

❏ Foundation/cracking problems

❏ Interior slab-on-grade problems

❏ Exterior concrete problems

❏ Fire damage problems

5—Electrical System

5.1 Service Entry   ❏ Replace   ❏ Retain

Capacity from street   Amps: Volts:

Overhead wire clearance

❏ Electric meter adequate

❏ Service entrance conductor adequate

5.2 Main Panelboard   ❏ Replace   ❏ Retain

Main circuit breaker   Amps: Volts:

Grounded to

❏ 15 Amp fuses/circuit breakers   No.

❏ 20 Amp fuses/circuit breakers   No.

❏ 25 Amp fuses/circuit breakers   No.

❏ 30 Amp fuses/circuit breakers   No.

❏ 40 Amp fuses/circuit breakers   No.

❏ Overcurrent protection adequate

Appendix E—Inspection Record

E-17

| | Data | | | Condition/needed repairs |

5.3 Branch Circuits   ❑ Replace   ❑ Retain

Circuit No.	Wire Gage	Wire Capacity	Area Served

❑ Circuits grounded to panel box
❑ Wire insulation in good condition
❑ Aluminum wire used

6—Plumbing

| Data | Condition/needed repairs |

6.1 Water Service Entry ❏ Replace ❏ Retain

Curb valve location

Line size   Material

❏ Shutoff valve operable

Water meter location

6.2 Interior Water Distribution Lines

❏ Replace ❏ Retain

Pipe size   Pipe material   Fixtures served

❏ Thermal protection adequate

6.3 DWV Piping ❏ Replace ❏ Retain

Pipe size   Pipe material   Fixtures served

❏ Vents, drains and traps operable

Data   Condition/needed repairs

6.4/6.5 Hot Water Heater   ❏ Replace   ❏ Retain

　Type　　　　　　　　　　Age

　Storage capacity　　　　　Gal.

　Recovery Rate

　❏ Plumbing components adequate

　❏ Fuel burning components adequate

　❏ Controls adequate

6.6 Water Well

　Location

　Depth of casing

　Pump type　　　　　　　Age

　Capacity　　　　　　　　GPM　　　　Depth

　❏ Pressure tank adequate

6.7 Septic system

　Location

　Tank capacity　　　　　　Gal.

　Age of system

　Size of drain field

　❏ Grease trap clean

6.8 Gas Supply in Seismic Regions

　❏ Service entrance has adequate clearance/flexible connection

　❏ Has automatic emergency shutoff valve

7—HVAC System

	Data		Condition/needed repairs

7.1 Thermostatic Controls ❏ Replace ❏ Retain

Location(s)

❏ Master switch operable

7.2–7.6 Heating System ❏ Replace ❏ Retain

Location

Fuel type

Fuel storage capacity

System type

Age of heating unit

BTU/hr output

❏ Room ventilation adequate

❏ Physical condition adequate

❏ Operation adequate

❏ Venting/draft adequate

❏ Distribution system adequate

❏ Controls adequate

7.7–7.10 Cooling System ❏ Replace ❏ Retain

Location

System type

Age of cooling unit

Electric service reqd.

❏ Physical condition adequate

❏ Operation adequate

7.11 Humidifier ❏ Replace ❏ Retain

Humidifier type

❏ Physical condition adequate

❏ Operation adequate

Appendix E—Inspection Record

	Data	Condition/needed repairs

7.12 Unit air conditioners ❏ Replace ❏ Retain

Location   Capacity in Tons   Volts   Amps

7.13 Whole House and Attic Fans

❏ Replace ❏ Retain

Location

Capacity in cubic ft/min.

Additional Notes

U.S. Department of Housing and Urban Development
HUD USER
P.O. Box 6091
Rockville, MD 20849

Official Business
Penalty for Private Use $300

FIRST CLASS MAIL
POSTAGE & FEES PAID
HUD
Permit No. G-795

February 2000

Made in the USA
Middletown, DE
25 February 2016